Peter Mvula, Meya Kalindekafe, Paul Kishindo,
Erling Berge, Friday Njaya (Eds.)

Towards Defragmenting the Management System of Lake Chilwa Basin, Malawi

Defragmenting African Resource Management

edited by

Associate Professor Dr. Mafaniso Hara
(University of the Western Cape)

and

Prof. Dr. Tobias Haller
(Universität Bern)

PLAAS
Institute for Poverty, Land and Agrarian Studies

A publication of the
Institute for Poverty, Land and Agrarian Studies (PLAAS)
University of the Western Cape
South Africa

Volume 1

ACP S&T PROGRAMME

LIT

Towards Defragmenting the Management System of Lake Chilwa Basin, Malawi

edited by

Peter Mvula, Meya Kalindekafe, Paul Kishindo,
Erling Berge, Friday Njaya

PLAAS
Institute for Poverty, Land and Agrarian Studies

LIT

This book is dedicated to all natural resources users within the Lake Chilwa basin

Cover Design: Friday Njaya
Front cover photos by Meya Kalindekafe

The editors would like to thank the European Union (project AFS/2009/219017) through the African Caribbean and Pacific (ACP) Science and Technology Programme (Europe Aid/127860/D/ACT/ACP) for the funding that made this book possible. The contents of this book though are our sole responsibility and do not reflect the views or position of the European Union.

Bibliographic information published by the Deutsche Nationalbibliothek
The Deutsche Nationalbibliothek lists this publication in the Deutsche Nationalbibliografie; detailed bibliographic data are available in the Internet at http://dnb.d-nb.de.

ISBN 978-3-643-90398-3

A catalogue record for this book is available from the British Library

Klosbachstr. 107
CH-8032 Zürich
Tel. +41 (0) 44-251 75 05
Fax +41 (0) 44-251 75 06
E-Mail: zuerich@lit-verlag.ch
http://www.lit-verlag.ch

LIT VERLAG Dr. W. Hopf
Berlin 2014
Fresnostr. 2
D-48159 Münster
Tel. +49 (0) 2 51-62 03 20
Fax +49 (0) 2 51-23 19 72
E-Mail: lit@lit-verlag.de
http://www.lit-verlag.de

Distribution:
In Germany: LIT Verlag Fresnostr. 2, D-48159 Münster
Tel. +49 (0) 2 51-620 32 22, Fax +49 (0) 2 51-922 60 99, E-mail: vertrieb@lit-verlag.de
In Austria: Medienlogistik Pichler-ÖBZ, e-mail: mlo@medien-logistik.at
In the UK: Global Book Marketing, e-mail: mo@centralbooks.com
In North America: International Specialized Book Services, e-mail: orders@isbs.com
e-books are available at www.litwebshop.de

Table of Contents

List of Figures

List of Tables

DARMA BOOK SERIES PREFACE

In Southern Africa the 'commons' have great livelihoods and biodiversity value, and they function as 'common pool resources' that are owned, accessed and used by multiple stakeholders in the region. However, as in many other parts of the world, management of the Southern African commons is fragmented. This fragmentation occurs in many ways and at different levels. For instance, authority is located in different government line agencies and technical expertise is spread among government agencies, non-governmental organisations (NGOs), universities and donors. In addition, actions take place at different geographic scales, and management networks are usually intra-sectoral (for example those in fisheries only know and network with others in fisheries). Funding for research and management also comes from multiple sources including foreign and international donors; and in some cases, commons transcend international borders, requiring international integration. Fragmentation in policy, legislation, knowledge base, authority and responsibility clearly makes a holistic approach to commons management difficult.

User stakeholders, especially the poor, usually draw upon the various resources (land, forests, fish, grass, water, wildlife) for composite benefits and livelihoods. If managed holistically, these commons can remain highly productive and sustainable. However, commons are in most instances under pressure due to their economic value and unco-ordinated management. Overexploitation and degradation occur particularly where ecosystem-based and ecosystem-wide institutional arrangements do not exist to overcome problems of subtractability that arise from open access.

In recent years, there have been increasing attempts to manage commons holistically as a result of the realisation and appreciation of the interrelatedness of both the ecological and socio-economic dynamics of natural resources – the so-called 'ecosystems approach'. This approach calls for partnerships between state agencies, research institutions, non-governmental organisations (NGOs), donors, traditional leaders, local communities and all other stakeholders. Effective links and collaborative partnerships among all stakeholders are critical for ensuring that the available scientific knowledge, indigenous knowledge, user knowledge and institutions are optimised. This book is an effort at demonstrating such a multi-disciplinary and inter-disciplinary approach in management of the Lake Chilwa Basin. It is based on contributions from various experts and users with long-term experience and knowledge of the area. The book reviews and evaluates policies and management approaches for the Lake Kariba environs, describes how these are fragmented and proposes a framework for defragmenting management.

The Lake Chilwa case study is one of seven cases under the Defragmenting African Resource Management project (DARMA) in Southern Africa, which aimed to build networks and research frameworks that could integrate resource management. The specific objective of DARMA was to increase national capacity in five African partner countries (Botswana, Malawi, South Africa, Zambia, Zimbabwe) for integrated management of commons with the overall broader objective being enhancement of biodiversity of exploited commons while alleviating poverty at the same time. The main strategy for achieving these objectives was the development of scientific networks and integrated knowledge bases that can combine different relevant natural and social science disciplines in a way that will be of practical use for both biodiversity conservation and sustainable development. The final outputs of the DARMA project are a series of books and journal articles that will serve as a valuable collection for environmental policy makers and will, hopefully, serve as catalysts in future debates on defragmentation of natural resources management in Southern Africa.

The preparation of this series of books has been made possible with European Union funding (Project AFS/2009/219017) through the African Caribbean and Pacific (ACP) Science and Technology Programme (Europe Aid/127860/D/ACT/ACP). The contents of these books are the sole responsibility of the authors. They do not reflect the views and position of the European Union. We thank all the stakeholders and participants who generously donated their time and energy sto making the DARMA project possible.

Development of the research frameworks for analysing fragmentation and defragmentation for the various DARMA case studies were developed jointly by the partners and associate partners. The DARMA partners and associate partners would particularly like to thank Professor Douglas Clyde Kongshøj Wilson (Innovative Fisheries Management, Aalborg University, Department of Development and Planning), an associate partner on the project, for all his contributions to the project.

Associate Professor Mafaniso Hara

DARMA Project Coordinator

Institute for Poverty, Land and Agrarian Studies

Faculty of Economic and Management Sciences

University of the Western Cape

FOREWORD

What is a commons? A commons is usually understood to be a piece of land (or more generally a flow of goods generated by natural processes) held jointly by a well-defined group of people larger than a household. Many commons are very productive and support large populations. However, all over the world these resources – including animals, birds, plants, forest resources and fish – are under pressure due to their economic value and unco-ordinated management. In this context, commonly shared resources suffer from degradation as each individual benefits fully from what he or she can subtract from the common pool, while the costs of the subtraction are borne by all users. And individual users have no incentive to limit their exploitation of resources if there is no institution to ensure that other users will also be limited. Gareth Hardin has described this unfortunate state of affairs as 'the tragedy of the commons'.

In Africa the knowledge base and management policy of the commons are fragmented: while the knowledge base is located in research institutes and universities, the commons are managed by government departments such as agriculture, forestry and wildlife. This fragmentation of knowledge and management makes a holistic approach to resource management of the commons difficult. Recently there have been attempts to manage the common through partnerships between state agencies on the one hand and traditional leaders, such as chiefs and village heads, and local communities on the other. However, biodiversity conservation efforts involving local communities have not always been successful. Economic pressure brought about by extreme poverty often militates against biodiversity conservation and sustainable use of common pool resources. Additionally, the dominance of a single-sector approach does not permit an appreciation of the interrelatedness of both the ecological and economic dynamics of natural resources.

It is against this background that the Centre for Social Research of the University of Malawi has collaborated with a number of research and government institutions in Botswana, Zimbabwe and Zambia to undertake a two-year project called Defragmenting African Resource Management (DARMA). The initiative is led by the Institute for Poverty, Land and Agrarian Studies (PLAAS) of the University of the Western Cape, South Africa. The project seeks to build research networks among natural and social scientists in universities and government departments to generate a knowledge base that can be applied to the sustainable utilisation and protection of biodiversity in southern Africa, as well as poverty alleviation. Integral to the project's broad objectives is conflict resolution because ecological degradation leads to competition for remaining resources. The final products of the project aimed at policy makers will include conflict mitigation strategies and incentive regimes for integrated resource management.

The common argument running through the contributions in this volume is that the resources of the Lake Chilwa basin are at risk if effective management structures and systems are not soon put in place. Declining resources are likely to engender further competition and conflict over the remaining resources. The contributing authors all agree that the only way forward is to defragment the area's management system.

Prof. Paul Kishindo
Chancellor College, Univeristy of Malawi

NOTES ON CONTRIBUTORS

Peter M. Mvula is a Senior Research Fellow with the Centre for Social Research at Chancellor College of the University of Malawi. He has a PhD in Rural Development from the University of East Anglia, United Kingdom (UK). He has done research and published in the fields of rural livelihoods, nutrition and food security, safety nets and the management and use of resources. His current research interests include rural livelihoods, safety and poverty.

Friday Njaya is currently heading the Planning Unit of the Department of Fisheries in Malawi. He has practical and theoretical aspects of fisheries governance, especially for the small-scale sector. He has published several papers, mainly on fisheries co-management. He has been a research partner for various projects co-ordinated by Institute for Fisheries Management (IFM), the World Fish Centre, the Food and Agriculture Organisation (FAO), the Christian Michelsen Institute (CMI) and the Institute for Poverty, Land and Agrarian Studies (PLAAS). He is also a member of the International Institute of Fisheries Economics and Trade (IIFET) and the International Association for the Study of the Commons (IASC).

Meya Patricia Kalindekafe is a Senior Lecturer in Ecology and an Environmental Impact Assessment Expert at Chancellor College at the University of Malawi. Her main area of research interest is ecosystems functioning. She holds a Bachelor of Science degree from the University of Malawi and a Masters in Ecology from the University of Wales, UK. Meya is a trustee of the Women in Science and Technology Network (WISTNET) in Malawi and an International Fellow of the Leadership for Environment and Development (LEAD).

Daimon Kambewa is a Senior Lecturer of Agricultural Extension in the Extension Department at Bunda College at the Lilongwe University of Agriculture and Natural Resources. Daimon has published research on customary land tenure in Malawi, the impact of HIV and AIDS on livelihoods, and social and power relations in the management of wetlands. The most recent publication by Daimon is *Access and Control in the Lake Chilwa Wetlands*, published in 2012 by LAP LAMBERT Academic Publishing GmbH and Co., Germany.

Wapulumuka Mulwafu is Associate Professor of Environmental History and Southern Africa Development Community (SADC) WaterNet Professorial Chair of Integrated Water Resource management at Chancellor College, University of Malawi. He has published extensively on irrigation reform, soil conservation and the socio-economic dimensions of water resource management. His most recent publication is *Conservation Song: A History of Peasant-State Relations and the Environment in Malawi, 1860-2000*, published in 2011 by the White Horse Press, UK.

Blessings Chinsinga holds an MPhil in Development Studies from the University of Cambridge, UK, and a PhD in Development Studies from the University of Mainz, Germany. He is an experienced academic, researcher and consultant and is currently based at the Department of Political and Administrative Studies, Chancellor College, University of Malawi as an Associate Professor specialising in Institutions and Development, Public Policy Analysis, Development Administration, Rural Livelihoods and Local Level Politics. He has published over 25 journal articles and book chapters including a book based on his PhD dissertation entitled *Decentralisation, Democracy and Poverty Reduction in Malawi*, published in 2007 by Köppe, Germany.

Milton Kutengule is an economist and independent consultant. He holds Doctorate in Philosophy, focusing on sources of income and economic growth, from the University of East Anglia, UK. He is also an expert in a range of subjects: public financial management reforms; fiscal and monetary policy analysis and management; projects design, management and monitoring and evaluation; economic modelling; decentralisation; poverty analysis and programming; population and development; social protection; social research; and rural livelihoods.

ACKNOWLEDGEMENTS

This book was produced with support from the European Union (EU) through the African Caribbean and Pacific (ACP) Programme (9th European Development Fund Lot 1), which funded the Defragmenting African Resource management (DARMA) project. The project was co-ordinated by Associate Professor Mafaniso Hara of the Programme of Poverty, Land and Agrarian Studies (PLAAS), based at the University of the Western Cape in South Africa. We wish to express our sincere thanks to Professor Hara for his guidance and support in reviewing and publishing this book. We would also like to sincerely express our gratitude to Professor Paul Kishindo and Erling Berge for editing this volume, as well as extend our thanks to the authors and various others stakeholders who contributed the papers that made this volume under the DARMA series possible. We further acknowledge Jonathan Gwaligwali for his expertise in producing the maps used in this volume.

Chapter 1

FRAGMENTED MANAGEMENT OF THE LAKE CHILWA BASIN

Friday NJAYA, PETER MVULA AND MEYA KALINDEKAFE

1.1 INTRODUCTION

The discourse on fragmented and decentralised commons in various ecosystems continues to generate interest in social and biological research aimed at finding ways of addressing conflicts in sustainable resource use and management. This book reviews research and published materials on the Lake Chilwa basin[1].The Lake Chilwa basin (see Figure 1) is a complex ecosystem that periodically experiences significant changes in water level, with negative consequences on the livelihoods of the local communities. On the basis of the reviews in this book, we offer strategies and recommendations in favour of a decentralised and defragmented management system that will help ensure sustainable use of the resources in the basin.

Lake Chilwa is a highly variable ecosystem that lies at 624 m above sea level between 35°45' E and 15°15' S in the centre of the low-lying Chilwa-Phalombe plain in the southern part of Malawi (FAO 1966; Ratcliffe 1971a). The size of the lake varies according to season but its maximum length is 38.5 km while its maximum width is 27 km (FAO 1966). In years of high water levels, the open water increases to 1,054 km^2, while in low water level conditions the open water recedes to 678 km^2.

The Lake Chilwa basin offers a wide range of natural resources for exploitation, such as fish, water, birds, small animals, pasture, land, timber, fuel wood, grasses (grazing), mushrooms and wild fruits. However, only some of these resources are managed in some way, namely fish, birds, water, land, timber, and fuel wood. Schuijt (1999) estimated an annual gross value of the wetland at US$21 million, which is derived from utilisation of the commons, with the largest contribution (82%) coming from fish, based on annual production estimated at 12,500 tons. The open lake and the basin contribute significantly to the livelihoods of the population of the catchment area, estimated at 1.6 million with a density of 321 persons per km^2, one of the highest in Malawi (NSO 2008). This high population density puts tremendous pressure on the area's resources.

---

[1] In this volume, the term 'basin' includes wetlands and/or floodplains, depending on the specific issue being analysed or discussed.

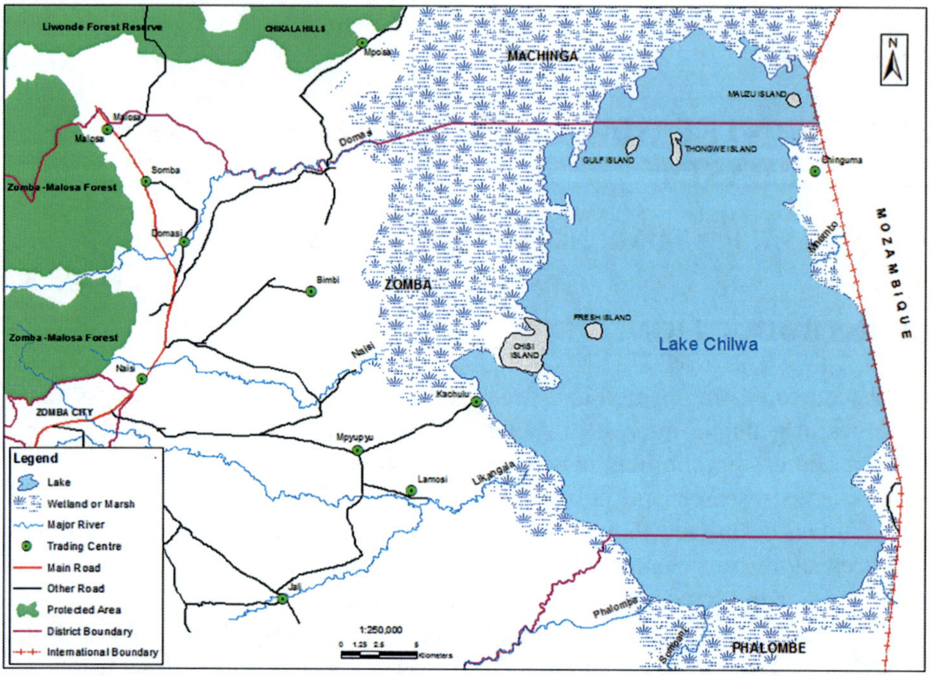

Figure 1: Map of Lake Chilwa basin, Source: NSO/DARMA Project

The definition of a common pool resource is based on the characteristics of *excludability* from harvesting and *rivalry* in consumption (Ostrom 2005). For example, to exclude any particular boat from harvesting fish in an area like this is extremely difficult, even more so when different countries are involved. And rivalry in the consumption of fish is illustrated by the fact that fish caught by one boat are not available for other boats. The same logic applies to other resources like water, birds, small animals, mushrooms, wild fruits and grasses (grazing). But for other resources in the area, such as land, timber, fuel wood and pasture, and with the current technology of exclusion, the logic of common pool resources is not obviously present. Exclusion is feasible. The problem with exclusion is not its cost but its distributional consequences, and these consequences are exacerbated by the interdependence of the resources.

One may regard the fragmented management of these resources as an attempt to introduce excludability for specific resources. However, the failure of this approach is due to the inherent non-excludability of some of the resources and the interdependence of all of them.

Fragmentation in commons management is thus a challenge often resulting in resource degradation and loss of biodiversity and limiting livelihood opportunities within the Lake Chilwa basin. Despite introduction of institutional arrangements and governance reforms that have advocated community participation within the basin since the 1990s, it is clear that management of the resources in the basin is fragmented, causing conflict amongst the managers of various resources as the papers in this book will show. Fragmented management is further compounded by the existence of numerous programmes and sectoral policies.

There is an emerging interest to identify strategies that focus on the protection of biodiversity of exploited commons and their use for poverty reduction through the development of scientific networks for integrated commons management. These networks include natural and social science disciplines and have links to policy makers and local people dependent on the commons. This book addresses pertinent issues such as how to prevent resource over-exploitation, foster economic development and implement conflict management.

Management of the commons in Malawi during the pre-colonial era (before the 1890s) was the responsibility of traditional leaders. In this period the commons were already centralised and defragmented (Chirwa 1996, Mvula 2002, Njaya 2009). During the post-colonial era (after the 1890s), the commons were managed by government agencies that focused on a particular sector (e.g. fisheries, water, land or forestry), using knowledge from scientists who were specialised in that sector and who sometimes worked with community groups to create and enforce rules developed for the sector. Once again management was centralised and defragmented. This scenario continued during the one-party state government in the post-independence era (1964-1994). The emergence of multiparty democracy in 1994 brought about a change in the management of the commons with a focus on community participation, which was decentralised but still fragmented since each sector worked with the same community using different strategies.

Fragmentation of management prevalent in the basin has led to unsustainable exploitation of the natural resources, which include fish, birds, water, land and soil. Therefore, a case can be made for the adoption of an ecosystem-based approach to natural resource management, which is based on a growing appreciation of the interrelatedness of the ecological and socio-economic dynamics of natural resources. In addition, new cross-disciplinary scientific networks are needed to identify areas of sectoral interconnectedness, to articulate and implement the necessary research and to translate the results into advice for policy makers.

To address the issue of fragmentation in the basin, the Defragmenting African Resource Management (DARMA) project was conceived and launched in 2009. Fragmentation of the knowledge base and, hence, of management policy in the Lake Chilwa basin commons were key areas of the project's research. DARMA built networks and research frameworks that could integrate natural resource management according to four dimensions:

- *technical expertise of natural resource management* within government management agencies, non-governmental organisations (NGOs), and universities;
- *natural resource management actions* across local, regional, national and international scales;

– *natural resource management powers, authorities and responsibilities* across different agencies and, in many cases, across national borders; and
– *natural resource management networks* across sectors.

The main objective of this book is to provide a review of the fragmentation of resource management in the Lake Chilwa basin. The review is based on the hypothesis that under fragmentation decentralisation, resource use is more sustainable than under defragmented centralisation, as it leaves more action on the local level.

1.2 RESEARCH FRAMEWORK

To assess fragmentation of the management system in the Lake Chilwa basin, researchers conducted a review from 2009 to 2012. In the review, the researchers applied the Drivers-Pressures-State-Impacts-Responses (DPSIR) framework as shown in Figure 2. The framework illustrates cause-and-effect relationships that can be represented through indicators. Several researchers have applied the DPSIR framework to various situations, but mostly to socio-ecological situations that centre on the social-ecological system (SES). In this review, we used the approach outlined by Carr *et al.* (2007), as it gave details about the genesis of the framework and how it fits in various socio-ecological situations.

As described by Carr *et al.* the *Drivers*, or Driving Forces, include socio-economic and socio-cultural forces driving human activities, which increase or mitigate pressures on the environment (e.g. human population increase and level of consumption). The *Pressures* relate to stresses that human activities place on the environment (e.g. use of chemical fertilisers) while *State of the Environment* refers to the condition of the environment (e.g. change in biodiversity). The *Impacts* are effects of environmental degradation (e.g. declining food availability and increased vulnerability to diseases). The *Responses* refer to responses by society to the environmental situation (e.g. formal policies and informal coping mechanisms).

To produce the chapters in this volume, a process was followed whereby scientists in various disciplines who had an interest in the management of Lake Chilwa basin were identified, after which these these scientists formed a network and then conducted review meetings and literature reviews. They collected both secondary and primary data. Some researchers collected primary data by using open-ended questions in interviews to gather information on current issues and conflicts, while others reviewed literature from previous studies to analyse the situation and identify conflicts. The researchers or scientists came from government, non-governmental organisations, academic and the private sector. At a meeting held in May 2011, participants discussed management of the natural resources in Lake Chilwa basin by assessing the status of the ecosystem, identifying key commons and conflicts in resource use patterns, and identifying governance reforms including tenure rights and prevailing policies. This meeting gave birth to a net-

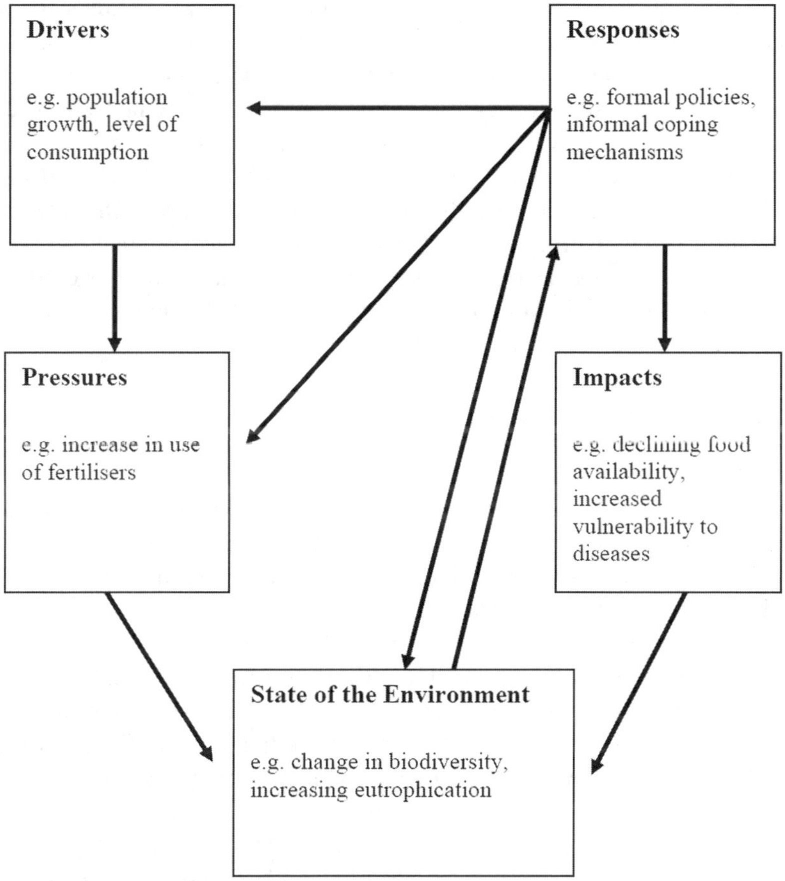

Figure 2: DPSIR Framework, Source: Adapted from Carr et al. 2007

work of the scientists who had done research or had on-going research related to the area. At the meeting, participants also outlined some previous and on-going projects that centre on the management of the natural resources in the basin. They adopted the DPSIR framework to guide the reviews included in this book.

Based on the DPSIR framework, the following three research questions guided the literature reviews:

- What is the current socio-ecological status of Lake Chilwa basin?
- For each particular discipline, what are the key conflicts in the resource use?
- What are the potential research areas for action research?

1.3 ANALYSIS

We analysed the fragmentation of resource management of the basin based on the reviews done by the various scientists who have an interest in the wetlands and have done research there over the past 20 years. We chose the areas for review through a process of first identifying the resources, and then the users, and finally the managers. Thereafter, we developed a matrix showing the resources and who had interest in a particular resource as a user or manager, which allowed us to see where overlaps could occur. We assessed the degree of influence of a particular entity on a particular resource: level 1 (***X***) indicated the least influence, while level 3 (***XXX***) indicated a strong influence. A blank space meant there was no influence by any policy or strategy. From that analysis, we ascertained that fisheries, birds, fuel-wood, land and water were key resources. These findings could not be analysed properly with without contextualising them in the historical setting, the ecosystem and the socio-economic spheres.

1.4 ORGANISATION OF THE BOOK

This volume contains papers that review commons in the basin. Specifically, the reviews centre on status, usage and management of the commons and also identify conflicts that arise as a result of both management and usage, which consequently impact on the livelihoods of the inhabitants. Health (including HIV and AIDS) and gender are cross-cutting issues in the reviews.

This book consists of ten chapters. This chapter (Chapter 1) provides background information; thereafter, Chapter 2 describes the ecological and environmental status of the Lake Chilwa basin. Chapter 3 analyses the migration order of the inhabitants of the Lake Chilwa basin and its implications for the control of and access to resources, followed by Chapter 4, which discusses the challenges facing the fisheries sector in its efforts to control overfishing on Lake Chilwa before Chapter 5 looks at the use rights under customary land tenure. Chapter 6 examines the basis of contestations over water use, while Chapter 7 re-examines the governance of the Lake Chilwa basin from an institutional perspective. Chapter 8 analyses the potential contribution of the Lake Chilwa basin to the local and Malawi economy. Chapter 9 presents an analysis of the issues highlighted by various researchers and the final chapter, Chapter 10, provides some strategies and recommendations for a defragmented and decentralised management of the commons in the Lake Chilwa basin, and the conclusion.

REFERENCES

Carr E, Wingard PM, Yorty SC, Thompson MC, Jensen NK & Roberson J. 2007. 'Applying DPSIR to sustainable development', *International Journal of Sustainable Development and World Ecology* 14: 543-555.

Chirwa WC. 1996. 'Fishing rights, ecology and conservation along southern Lake Malawi, 1920-1964', *African Affairs* 95: 351-377.

Food and Agriculture Organisation (FAO). 1966. *Report to the Government of Malawi: A Programme for Fisheries Development.* Based on the work of HLF, Renson,

UNDP/TA Fisheries Development Adviser, FAO No. TA2239. United Nations Development Programme FAO, Rome.

Krebs CJ. 2009. *Ecology: The Experimental Analysis of Distribution and Abundance.* 2nd Edition. Harper and Row, New York.

Mvula P & Haller T. 2009. 'Common pool resource management in Lake Chilwa, Malawi: A wetland under pressure', *Development Southern Africa* 26(4): 539-553.

Njaya FJ. 2009. 'The Lake Chilwa fishing household strategies in response to water level changes: migration, conflicts and co-management.' *PhD thesis*. University of the Western Cape, South Africa.

National Statistical Office (NSO). 2008. *Malawi Population and Housing Census.* NSO, Zomba, Malawi.

Ostrom E. 2005. *Understanding Institutional Diversity.* Princeton University Press, United States.

Pomeroy RS, Rivera-Guieb R. 2005. *Fishery Co-management: A Practical Handbook.* International Development Research Centre, Canada.

Ratcliffe C. 1971. 'The fishery of the lower Shire area, Malawi', *Fisheries Bulletin* 3.Ministry of Agriculture, Zomba, Malawi.

Schuijt KD. 1999. *Economic Valuation of the Lake Chilwa Wetland.* State of the Environment Study No. 11. Lake Chilwa Wetland Project, Zomba, Malawi.

Chapter 2

LAKE CHILWA BASIN ENVIRONMENTAL STATUS IN RELATION TO BIODIVERSITY, POLLUTION AND HABITAT CHANGE

Meya KALINDEKAFE

2.1 INTRODUCTION

The Lake Chilwa basin refers to the *watershed* or *catchment area* or land (including streams and rivers) where all water from the rain drains downhill into the lake. There is vast open land including its resources within the Lake Chilwa basin that people use in common. However, these common resources – which include animals, birds, plants, forests resources, fish, land and water – are under pressure, largely due to their economic value and unco-ordinated management. Consequently, the biodiversity of the basin also risks being lost due to overexploitation and destruction of certain species.

In this chapter, I identify sources of pollution, siltation, biodiversity and habitat change in the Lake Chilwa basin through a review of existing literature. On the basis of this critical review, I recommend areas for action research which would strengthen the management of the common resources.

2.2 DEFINITIONS USED IN THIS CHAPTER

Krebs (2009) defines *ecology* as the scientific study of the *interactions* that determine the *distribution* and *abundance* of organisms. Malawi's Environment Management Act (EMA) of 1996 defines the *environment* as the physical factors of the surroundings of the human being including land, water, atmosphere, climate, sound, odour and taste, as well as the biological factors of fauna and flora. The definition includes the cultural, social and economic aspects of human activity, and the natural and built environment. *Biodiversity* refers to all aspects of variety in the living world, namely the variations between ecosystems (aquatic or terrestrial), between habitats, between different species and between individuals in a species. *Pollution* is the introduction (by natural causes or human activity) of a contaminant into the environment, making the water, air and land dirty, and thereby causing instability, disorder, harm or discomfort to the ecosystem. *Siltation,* on the other hand, occurs when silt becomes stationary in a particular area and begins to obstruct or choke out the normal flow and life in a water body.

2.3 LOCATION OF THE LAKE CHILWA BASIN

Lake Chilwa is a tropical lake located in Southern Malawi, to the north of the Phalombe Plain between Zomba and Mulanje (Figure 1). It has no outlet. Its catchment area consists of the Phalombe, Zomba and Machinga Districts and part of Mozambique on the eastern side. The catchment area has a total of 8 349 km^2, with 5 669 km^2 in Malawi and 2 680 km^2 in Mozambique. A total of 233 km^2 is used for wetland cultivation.

The Lake Chilwa basin has a diversity of natural habitats and land users. The habitat types found in the wetland include marshes, swamps dominated by *Typha domingesis*, rivers, open water lake, island and floodplain. In normal years, one third of the lake is open water, one third is swamp and another third is floodplain. Free floating species such as *Pistia stratiotes* and *Ceratophyllum demersus* are found on the lake edge of the swamp. The large sedge *Scirpus littoralis* and the aquatic grass *Paspalidium germinatum* commonly occur on open water. The vegetation of Lake Chilwa is influenced by the seasonal fluctuation of water levels such that in drier years some species disappear. Human activities such as farming contribute to the lake's fluctuation through siltation.

2.4 CLIMATE

Rainfall distribution in the lake's catchment is uneven. In the highlands rainfall ranges between 1 100 to 1 600 mm per year, reaching 2 000 mm in Zomba and Mulanje Plateaus. The lowland plains receive much less rainfall of 800 to 900 mm per year. Rain falls in the months of November to April followed by a long dry season in the months of May to October. In years of heavy rainfall floods are common in some parts of the Lake Chilwa basin. Mean annual precipitation is 85.6 mm, recorded at an elevation of 1 029 m.

Temperatures in the Lake Chilwa catchment vary with altitude. Average temperature of the warmest month is 30°C and the average temperature of the coldest month is 16°C.

2.5 CURRENT ECOLOGICAL STATUS OF THE LAKE CHILWA BASIN

People living around Lake Chilwa derive their livelihood from utilisation of the wetland, directly or indirectly. The resources they use from the wetland include fish and birds for sale and consumption, as well as building materials for houses and boats, brooms, mats, fish and bird traps and baskets. They also use water for irrigation, transport and domestic consumption. The benefits which the local population derives from the wetland make it an immensely valuable asset (Government of Malawi 2000). The most valuable natural resources under heavy exploitation in the Lake Chilwa are birds and fish. Lake Chilwa and its wetlands are neither a National Park nor a Protected Area so people have free access to the natural resources. Their uncontrolled use of natural resources puts great pressure on and threatens the sustainability of the natural resources. The population pressure in the wetland is high, with 916 477 people living in the catchment area and 77 000 people living in the wetland itself. Average population density is 164 persons per km^2. According to Kalanda-Sabola (2007) historical trends indicate that, while some of the resources in the wetland have declined, others have maintained their status. Forests, fish, and typha reeds are declining while hippos and birds have more or less maintained their status.

At 164 persons per km^2 the population density of the Lake Chilwa catchment area is one of the highest in the country. The population of the area largely depends on natural resources such as land, water, trees, birds, rodents and others for survival. An estimated 1.2 million birds are trapped every year, involving 461 bird trappers (Maloya 2001). Birds are an important source of protein as well as income. The lake contributes between 25% and 30% of the total fish caught in Malawi and economically sustains an estimated 6,000 fishers.

Owing to the high human population the area's natural resources, especially birds are overexploited. The waterfowl is heavily utilised for consumption. At the beginning of the 20th century it was reported that hippos, kudus, sable, antelope, bushbuck, reedbuck, hartebeest, lions, hyenas and leopards were plentiful in the Lake Chilwa basin. These are hardly visible now. The decline of large mammals in the lake basin had been reported by many researchers (Sweeney 1965, Howard-William 1973). Murray (1930) recorded only warthogs, bush pigs, baboons, bushbucks, lions, leopards and hyenas while an extensive survey by Sweeney (1965) recorded bushbucks, reedbucks and duikers. None of these mammals were reported in the studies in 1990s, showing that such species were becoming rare. Mfune and Mhango (1998) found that most of the mammals were in the small mammal group such as a spiny mouse, four-toed hedgehog and a field rat.The decline of big mammals is attributed to a number of causes, including change of the environment, destruction of the mammalian fauna due to periodic recession of the lake, hunting, and habitat destruction through cultivation, firewood, the construction of dug-out canoes and human settlement.

The most valuable natural resources under heavy exploitation in the Lake Chilwa basin are plants, water, birds and fish. In 1997 and 2006, the lake was declared a Ramsar Site and a Man and Biosphere (MAB) reserve, respectively. With regard to its status as a Ramsar Site, people still have free access to the resources but based on the principle of 'wise use' of natural resources, while its status as an MAB Reserve means that people have limited or controlled access to some parts of the lake and its wetland. As an MAB reserve the lake is divided into three management zones: the *core zone*, which is the area of minimal activity, the *buffer zone*, where only limited activities compatible with the conservation objectives can take place, and the *transition zone*, which includes areas of economic activity where sustainable resource management practices are promoted and developed. However, because of poor monitoring for compliance, as well as other pressures such as poverty, increasing numbers of people have easy and unlimited access to the common resources (Department of Biology 2006).

2.5.1 Amphibians

There are several factors that threaten the population of amphibians in the Lake Chilwa basin (Mfune and Mhango 1998). Firstly, is the utilisation of small frogs and toads as fishing bait. Mfune and Mhango (*ibid*) found that this practice was common around Swang'oma and Chisi Island. Secondly, amphibians are sensitive to prolonged droughts because their breeding sites dry up. Thirdly, Lake Chilwa is also prone to being affected by climate change and this further threatens amphibian life. Thirdly, amphibians are preyed on by birds and snakes, which reduces their numbers. And finally, habitat destruction through cultivation is another threat to the amphibians.

2.5.2 Reptiles

The reptiles found in the basin include aquatic, amphibious, fossorial and open ground species (Kabwazi and Gulule 1999). About 19 species were recorded by Stevens (1974) and four species were recorded by Mfune and Mhango (1998). However in the subsequent studies, most of the species that were recorded are snakes. It is suggested that habitat changes due to human settlement indirectly favour snakes. According to Morris (1993), as cited by Kabwazi and Gulule (1999), agricultural practices favour mice, which are food for snakes.

2.5.3 Invertebrates

Invertebrates are the largest group of wild animals in the Lake Chilwa basin (Kalk *et al.* 1979). Three hundred-and-eleven species of invertebrates have been recorded in Lake Chilwa, which is believed to be only a fraction of the species that occur in the basin. Insecta is the largest class of the invertebrates in the Lake Chilwa basin with 12 orders, 169 genera and more than 254 species, and the order Lepidoptera with 60 species is the most abundant followed by Hymenoptera with 43 species (Kabwazi and Gulule 1999).

2.5.4 Flora

The Lake Chilwa wetland is located within the deciduous forest region, marked by dry and wet seasons. There are almost no forests on customary land except small patches in graveyards. An increase in the fishing industry has created more demand for wood for processing fish and making dugout canoes. The fertility of the wetland and its catchment has also promoted agricultural activities – such as maize, rice, cassava, tobacco and vegetable farming – that sustain thousands of people in the catchment and beyond. Agriculture has led to massive clearing of forests and in 2000 the deforestation rate was estimated at 3.2% per annum (Government of Malawi 2000). Forest patches that exist now harbour remnants of the original vegetation of the catchment area, except the Zomba Forest Reserve where most of the indigenous vegetation was cleared and replaced with exotic trees species (*Pinus patula).*

2.5.5 Birds

Lake Chilwa is particularly rich in bird species; about 164 bird species are specifically associated with the Lake Chilwa wetland, congregating there for feeding, resting and breeding (UNESCO 2011). Forty-one of these species are palearctic migrants and 14 are intra-African migrants. These migrant birds are heavily hunted for human consumption. Unfortunately, the nesting period of many of these birds coincides with the hunger period experienced by the local people at the end of the rainy season. The following bird species could thus be vulnerable: the fulvous whistling-duck (*Dendrocygna bicolour*), the lesser moorhen (*Gallinula angulata*), the lesser gallinule (*Porphyrio alleni*), the black-crake (*Amaurornis flavirostris*), the grey-headed gull (*Larus cirrocephalus*) glossy ibis (*Plegadis falcinellus*), the spur-winged geese (*Plectropterus gambensis*), the African spoonbills (*Platalea alba*), the black-crowned night-herons (*Nycticorax nycticorax*) and the African skimmers (*Rynchops flavirostris*).

Many species of birds breed during the months January to July at various sites around the lake and it is also a time when migrant bird species from many palearctic visit the lake (Wilson 1999). Most birds favour the river mouths for breeding, as these areas are largely inaccessible to the local people and offer protection to the breeding birds because of the dense marsh vegetation, (Bhima 2006). Coincidentally these are also the core zones under the Man and Biosphere Reserve status.

Local hunters trap a range of species in large numbers such as the common moorhen *Gallinula chloropus*, lesser moorhen, Allen's gallinule, Blake-crake, fulvous whistling-duck, white-faced whistling-duck *Dendrocygna viduata* and Hottentot teal *Anas hottentota* (Wilson and van Zegeren 1998, Wilson 1999). There are 461 bird trappers operating along Lake Chilwa's beaches (Government of Malawi 2000). When the water level drops, their catch of birds greatly increases by about 300 to 500% (Bhima 2006).

In the open water, predators such as the resident pink-backed pelican *Pelecanus rufescens*, the grey-headed gull *Larus cirrocephalus* and the migrant white-winged black tern *Chlidonias leucoptera* are common, especially in Kachulu Bay, a major fishing centre. Birds of prey found at the Lake Chilwa wetland include the African marsh harrier *Circus aeruginosa* and the much less common fish eagle *Haliaetus ranivorus*. The yellow-billed kite *Milvus aegypticus* and the lesser kestrel *Falco naumannii* represent the palearctic migrant birds of prey in the Lake Chilwa wetland (Maloya 2001, Bhima 2006).

2.6 THREATS TO THE BIODIVERSITY OF LAKE CHILWA BASIN

There are many threats to the biodiversity of Lake Chilwa basin and its catchment, which have an impact on the overall lake basin as they create an imbalance in the ecosystem. Furthermore, the continued population pressure and unsustainable ways in which natural resource are utilised, put the basin in danger of ecosystems collapse. The threat emanates from various sources or a combination thereof:

- Humans destroy habitat by opening vegetable gardens and building houses.
- Deforestation poses a threat to wildlife as habitats get destroyed.
- Animals such as birds become an alternative source of meat when the lake dries up, when the water level drops, and sometimes when the fishing season is closed, thus putting great pressure on their utilisation.
- Untargeted hunting leads to catching any animal found in the process.
- Uncontrollable bush fires destroy the biodiversity.
- Hunting occurs during the off-fishing season as an occupation.
- Some people hunt certain animals for elimination because they are regarded as a nuisance in the area, for example birds that damage rice paddies.
- Natural disasters such as floods and droughts lead to a decline in species when they occur. In the past when water levels receded, there is significant evidence that great number of species were lost.
- No fish species is recorded as threatened in Lake Chilwa. However, intensified fishing and periodic drying up of the lake could jeopardise those fish species that constitute a viable commercial fishery.

2.6.1 Water pollution

With regard to water resources, factors leading to changes in biodiversity include siltation, damming, warming, pollution and changes in hydrological regimes, which cause reproductive failure. Most of the changes in the watershed are related to human population pressure and poverty, which have resulted in the unsustainable use of water, forestry and fishery resources. In Lake Chilwa, fish mortality has been reported to be influenced by low levels of dissolved oxygen, high silt load, high alkalinity, low water levels and high pH (Msiska 2001).

Jamu *et al.* (2006) reveal a range of pollution sources in the Lake Chilwa basin: estate agriculture, which introduces pesticides and fertilisers; smallholder agriculture, which brings topsoil and household wastes; deforestation, which encourages the erosion of topsoil and organic matter; irrigation schemes, which contribute pesticides and fertilisers; and urban centres, which release organic and inorganic pollutants. However, studies have not indicated which one among them is the greatest polluter. Future research should also help to establish how much pollution each river contributes to the Lake Chilwa, taking into account the different activities happening in the rivers' catchment. This would help stakeholders to determine the magnitude of the damage being done to the lake and identify specific intervention measures to solve the problem.

With the increasing population pressure and agricultural activities in the catchment, these pollutants levels are likely to increase. A study by Jamu *et al.* (2006) on the Likangala and Domasi rivers – which empty their waters into the lake – implicated river flow, conductivity and levels of total suspended solids (TSS) in the water as influences on upstream migration of *Barbus* species (*matemba* fish). The researchers also observed that increasing deforestation and the conversion of agricultural land to homestead development have led to erosion and sedimentation, particularly in the Likangala catchment basin. Lake Chilwa receives water from the Domasi, Likangala, Namadzi, Sombani, Thondwe and Phalombe rivers, whose waters are being used for different purposes such as agricultural production.

Furthermore, the increasing use of water for industrial and other productive activities in the upstream area of Likangala River creates numerous problems for downstream communities. The establishment of an abattoir near the Likangala Bridge and sewage leaks from Zomba Central Prison, Zomba Central Hospital, Cobbe Barracks and Zomba Mental Hospital have resulted in the dumping of waste into the river (Mulwafu 2000). These activities all contribute to the pollution of the water of the river.

2.6.2 Land use changes

Research in the Chilwa Basin documenting land use change has shown a net increase in cultivation, including a conversion of wetlands to rice production and settlement. Research in the early 2000s along the Likangala, Naisi and Domasi rivers revealed that there was intensification in use of stream-bank gardens (*dambo* cultivation) and wetland areas and there was an increase in competition and conflict over control and use (Peters 2004). However, recommendations on *dambo* protection and conservation of areas where cultivation is done have not been picked up by the authorities, despite the turn to environmentally conscious policies since the early 1990s. This oversight is witnessed by more recent decisions by government and donors that irrigation is central to achieving a more food secure population, with farmers turning their attention to *dambos* and wetland cultivation to increase production. Increased demand has put wetlands in the country under threat, including Lake Chilwa where rice cultivation has intensified.

2.6.3 Sand mining

The sand mining activity taking place at different points along the Likangala River not only pollutes the water but also contributes to the destruction of river banks, thereby altering the natural flow of the river. People downstream experience problems with the muddy water. In addition, settlements in the upstream area, such as Chikanda, St. Mary's and Mpondabwino, further contribute to the pollution of the river.

2.7 THE EFFECTS OF CLIMATIC CHANGES ON LAKE CHILWA BASIN'S ECOLOGY

Lake Chilwa is characterised by marked seasonal and annual fluctuations in water levels, which are at times amplified by periodic recessions, with ultimate drying up of the lake. Climate change is a major issue facing the people and the environment of Lake Chilwa, as climate induced risks and hazards can have wide ranging, often unanticipated, effects on the environment, biodiversity, water resources, energy, and terrestrial and aquatic ecosystems. These changes affect the distribution and dynamics of flora and fauna of the Lake Chilwa wetland. The effects, however, are not adverse as these periodic recessions, resulting in the drying and filling of the lake, are cyclical in nature. Lake Chilwa has experienced nine minor and moderately severe recessions of which the latest occurred in 1995 due to drought (Njaya 2001).

Recent years of drought have caused Chilwa to shrink, so that huge areas of lakebed swamps and sands now sit on land that used to be well submerged, but the lake still occupies a good 2 500 km^2. In 1859 the Scottish missionary explorer David Livingstone suggested that Chilwa reached almost to the surrounds of Mount Mulanje, about 30 km further south than its present limit. It would have been contiguous with Lake Chiuta, and about three times deeper than its present maximum depth of about 3 m.

During the minor and moderately severe recessions, fish mortalities and reduced catches were recorded (*ibid*). Rainfall in the basin is around 1 000 to 1 200 mm per year. However, during the dry season, this water is lost through evaporation processes. Since the lake is shallow, it is vulnerable to drying up during prolonged drought spells (Chavula 2000). Climate change is an issue that threatens the Lake Chilwa wetland such that prolonged droughts are likely to cause the lake to dry up, which would affect the entire biodiversity of the wetland and the community at large whose livelihoods depend on the lake. Another human activity within the lake basin that impacts on the environment is bush burning, which causes air pollution and also removes the carbon sink.

2.8 MANAGEMENT OF THE LAKE CHILWA BASIN

Over the past decades, several programmes and projects in the Lake Chilwa basin have been launched, aiming at conserving the resources and improve livelihood of the local communities. As these different development and research programmes have their own different objectives, management of the resources has been fragmented.

In 1996 Malawi developed the National Environmental Policy (NEP) framework, which aims at narrowing the gap between the degradation of natural resources and environment and promoting sustainable production and economic growth (Government of Malawi 2000). The policy was reviewed in 2006. It advocates for sustainable use of land, water and biological diversity of resources. In the case of Lake Chilwa wetland, the policy is in line with 'wise use' concept of natural resources as advocated by the Ramsar convention. In 1996, the Environmental Management Act (EMA) was passed. The Act addresses cross-sectoral issues and regulates environmental impact assessment, conservation of biodiversity, integrated pollution control, control of pesticides and hazardous substances. This Act is useful for environmental management and natural resources conservation.

The government has developed other policies such as the National Forestry Policy of 1996, the National Wildlife Policy of 2000, the National Fisheries Policy of 2001, the National Water Policy and the National Land Resource Management Policy of 1998. These policies have subsequently led to formulation of various acts that prescribe the dos and don'ts as far as natural resources conservation and management are concerned.

The open waters and the land of Lake Chilwa under government-irrigated rice cultivation are designated as public land under the control of the government. The rest of the land in the Lake Chilwa wetland is customary land, which is allocated to villagers and controlled by local chiefs. The various land and natural resource stakeholders in the Lake Chilwa wetland include: Government departments such as fisheries, wildlife, agriculture, water and rural community, and conservationists such as the Wildlife and Environment Society of Malawi, whose interests often conflict. By 2000 there was no established legal mechanism for the conservation and sustainable utilisation of natural resources of the Lake Chilwa wetland and there were no user rights or legal tenure system given to user groups (Maloya 2001).

The designation of Lake Chilwa as a Ramsar Site has attracted the attention of international agencies such as the Danish Development Agency (DANIDA), which has supported activities aimed at conserving the wetland and catchment area. The Lake Chilwa Wetland and Catchment Management Project, which was supported by DANIDA, helped in promoting a development approach that aims at assisting communities to help themselves by conserving and sustainably utilising natural resources in the wetland through active participation. The project has helped the government of Malawi meet its obligations in the Ramsar Convention by producing the Lake Chilwa State of the Environment Report in 2000, which formed the basis for development of a Lake Chilwa Management Plan in 2001. The plan is an action plan directed at fulfilling the main obligations of the Ramsar Convention, i.e. maintaining the ecological character of the region, with wise use and sustainable utilisation of the natural resources (Government of Malawi 2001, Jamu *et al.* 2006).

2.9 TRANSBOUNDARY MANAGEMENT CHALLENGES

The Lake Chilwa wetland is an ecologically and economically important area that is shared by Malawi and Mozambique. As a result, the wetland fishery is under threat from overfishing and siltation of breeding sites in influent rivers from the two countries (Jamu *et al.* 2006). It is therefore imperative that whatever management options for whatever resource are put in place should be done in consultation with officials from Mozambique. It has been done before (*ibid*) and it can be done again. Legal instruments such as the SADC Protocol on Fisheries and Shared Waters can be used in facilitating the transboundary management of the shared resources. Unfortunately, the two countries have different management regimes for some of the resources found in the Lake Chilwa basin, rendering integrated management difficult.

A study by Njaya (2001) sought to identify major challenges of implementing co-management of Lake Chiuta, a shared ecosystem between Malawi and Mozambique, and found a conflict between the two countries. The Malawian fishing community was represented by Beach Village Committees and claimed that the seining operations allowed in Mozambique were destroying habitats for fish breeding, creating serious conflict in managing the fisheries resources.

2.10 PROPOSED ACTION RESEARCH TO SUPPORT COMMONS MANAGEMENT

With the issues raised above, I have identified a clear need to carry out action research to improve the ecological and environmental status of the lake basin in relation to water pollution, siltation and climate change, which will entail developing a model of energy flow and nutrient and material cycling within the Lake Chilwa Basin. This action research will enable researchers and local communities and other key stakeholders to come up with a scenario where anybody can see how different components of the basin are connected to each other and appreciate the need for conserving the resources though sustainable exploitation. Proposed areas for action research are:

- determining the key sources of pollutants, quantifying them and monitoring how much each river is contributing towards water pollution and siltation;

- developing improved measures for conservation-based ecological farming, including sustainable soil and nutrient management practices;
- developing a long-term ecological programme for monitoring water quality of the rivers and lake in relation to habitat biodiversity; and
- determining how different gender groups access the resources in the basin and establish how this would affect the functioning of the lake basin.

2.11 CONCLUSION

In this chapter I set out to examine the critical ecosystems that relate to the exploitation of the commons in the Lake Chilwa basin by reviewing existing studies. Results from the review indicate that the area's natural resources are under threat largely due overexploitation and climate change. The destruction of these natural resources poses a threat to the livelihoods of large number of people. I urgently recommend that government and other stakeholders develop a framework for sustainable utilisation of the resources that is based on research evidence.

REFERENCES

Bhima R. 2006. 'Subsistence use of waterbirds at Lake Chilwa, Malawi' pp255-256, in: Boere GC, Galbraith CA & Stroud DA (eds) *Waterbirds Around the World*. The Stationery Office, Edinburgh, United Kingdom.

Chavula GMS. 2000. *The Evaluation of the Present and Potential Water Resource Management for the Lake Chilwa Basin*. US Land Tenure Centre, University of Wisconsin. Available from: http://wisc.edu/Itc/baspusafri.html (Accessed on 26 December, 2001).

Department of Biology. 2006. *Lake Chilwa Man and Biosphere Nomination Form*, Report submitted to UNESCO Malawi Office, Lilongwe, Malawi.

Government of Malawi. 2000. *Lake Chilwa Wetland: State of the Environment*. Environmental Affairs Department, Lilongwe, Malawi.

Howard-William C. 1973. 'Vegetation and environment in the marginal areas of a tropical African Lake (Lake Chilwa).' *PhD thesis*. University of London, United Kingdom.

Jamu DM, Chimphamba JB & Brummett RE. 2003. 'Land use and cover changes in the Likangala catchment of the Lake Chilwa basin, Malawi: Implications for managing a tropical wetland', *African Journal of Aquatic Science* 14: 119-132.

Kabwazi HH & Gulule MS. 1999. *An Inventory of Wildlife around Lake Chilwa with Special Emphasis on Threatened Species, Study No. 7*. Lake Chilwa Wetland State of the Environment, Zomba, Malawi.

Kalanda-Sabola MD. 2007. 'Assessing indigenous knowledge of forest and lake resources degradation: A case of Chisi Island, Lake Chilwa, Zomba, Malawi, Central Africa', *Journal of Agricultural, Food, and Environmental Science* 1(2):1-25.

Kalk M, McLachlan AJ & Howard-Williams C (eds). 1979. *Lake Chilwa: Studies of Change in a Tropical Ecosystem*. Monogralhiae Biologicae, The Hague Publishers, London.

Maloya H. 2001. ?*Community-Based Natural Resource Management: The case of Lake Chilwa Wetland Project, Malawi*. Zomba, Malawi.

Mfune JKE and Mhango CRG. 1998. *Amphibians, Reptiles and Mammals of Lake Chilwa: The Lake Chilwa Environment: A Report of the 1996 Ramsar Site Study.* Chancellor College Publications, Zomba, Malawi.

Msiska OV. 2001. 'A Review of Limnology of Malawi' pp121-189, in: Wetzel RG and Gopal B (eds) *Limnology in Developing Countries 3.* International Association for Limnology, New Delhi, India.

Mulwafu WO. 2000. *Conflicts over Water Use in Malawi: A Socio-Economic Study of Water Resource Management along the Likangala River in Zomba District.* From the 1st WARFSA/WaterNet Symposium: Sustainable Use of Water Resources, Maputo, 1-2 November 2000.

Murray SS. 1932. *A Handbook of Nyasaland.* Government Printers, Zomba, Malawi.

Njaya FJ. 2001. *Review of Management Measures for Lake Chilwa Malawi: Final Project Report.* United Nations University, Reykjavik, Iceland.

National Statistical Office (NSO). 2008. *Malawi Population and Housing Census.* Government of Malawi, Zomba, Malawi.

Peters P. 2004. *Stream-bank and Wetland Gardens.* BASIS CRSP Management Entity Department of Agricultural and Applied Economics, University of Wisconsin-Madison.

Stevens RA. 1974. 'An annotated checklist of the amphibians and reptiles known to occur in south-eastern Malawi', *Arnoldia* 6(30): 1-22.

Sweeney RCH. 1965. 'Notes on some birds seen in the Lake Chilwa region', *Society of Malawi Journal* 18(1): 55-58.

Van Zegeren K and Munyenyembe MP. 1998. *The Lake Chilwa Environment: A Report of the 1996 Ramsar Site Study.* Department of Biology, Chancellor College, Zomba, Malawi.

Wilson JGM and van Zegeren K. 1998. 'The birds of Lake Chilwa'pp 91-108, in: Van Zegeren K and Munyenyembe MP (eds) *The Lake Chilwa Environment: A report of the 1996 Ramsar Site study.* Department of Biology, Chancellor College, Zomba, Malawi.

Wilson JGM. 1999. *The Waterfowl of Lake Chilwa and their Utilisation by Local Communities and Conservation Measures as Required by the Ramsar Convention.* State of the Environment Study No. 20, Lake Chilwa Wetland Project. DANIDA/Chancellor College, Zomba, Malawi.

Chapter 3

HISTORICAL SETTLEMENT PATTERNS, CONTROL AND USE OF NATURAL RESOURCES IN THE LAKE CHILWA BASIN

Peter MVULA

3.1 INTRODUCTION

Up to the beginning of the 20th Century, the history of the Lake Chilwa Floodplain was dominated by major population migrations. Driven by circumstances in their original homelands, several ethnic groups came to settle in the area: first the Nyanjas, then the Yaos and later the Lomwes settled in the area around the lake. Pachai (1979) observes that the pre-colonial history of the floodplain was influenced by events that went beyond the rise and fall of its water levels. During the colonial and postcolonial period, economic opportunities through agriculture, fishing and fish trading attracted more people to the area. As a result of the continuous migrations into the area, society has become layered based on the order of settlement (Agnew 1972) and the order of settlement has implications for rights of access to resources as well as for their control.

Understanding the livelihood systems in the Lake Chilwa ecosystem, therefore, requires an understanding of the people who settled in the area, when they arrived and what had motivated them to come. In this chapter I review literature that sheds light on the migration patterns of the inhabitants of the Lake Chilwa Floodplains.

3.2 HISTORY AND ETHNOGRAPHY OF THE LAKE CHILWA BASIN

The Lake Chilwa basin covers an area of over 7 500 km^2. It occupies parts of three administrative districts in the southern region of Malawi, namely Zomba to the west, Machinga to the north, and Mulanje and Phalombe to the south. These areas come under the jurisdiction of eight chiefs: Mwambo, Mkumbila and Kumtumanje in Zomba; Kawinga, Mlomba and Mposa in Machinga; and Mkhumba and Nazombe in Phalombe (Agnew 1970 and 1972). The Nyanja who are believed to be the earliest inhabitants of the floodplains live all around the lake, and also have permanent villages on Chilwa Island. The Mpotola, who are believed to be related to the Lomwe, are found around the Chikala hills under the authority of sub-traditional authority Mposa. The Yao are scattered all over the whole floodplain and also occupy dry season villages on Chisi, Chenjelani and Tomwe islands, moving back to their mainland villages with the onset of the rains. The Lomwe are mainly found along the eastern side of the lake.

Although linguistically diverse, the three dominant groups are culturally similar (Chilibvumbo 1969; Kalk 1990). They are all matrilineal; they practise matrilineal inheritance, and uxorilocal marriages are the norm. Primary rights to land belong to women, with men enjoying only secondary rights for the duration of their marriage to a right-holding woman; children are deemed to belong to their mother's side.

Rain-fed agriculture is the dominant economic activity. Fishing and irrigated rice farming also occupy significant proportions of the population.

3.2.1 The Nyanja

All accounts point to the fact that the Nyanja were the earliest inhabitants of the Lake Chilwa basin. They were however, later dominated by the militarily superior Yao who arrived from the Mozambique side of the lake in the mid-19th Century. By the mid-17th Century, the Nyanja, an off-shoot of the Chewa, inhabited the area lying north of the Zambezi, east of the Luanga River in Zambia and west of the Shire River, to the southern and south-western shores of Lake Malawi (Alpers 1969).

The ancestors of the Chewa are believed to have originated from the current Democratic Republic of the Congo. The most accepted theory is that their arrival in the Lake Malawi region took place some time in the 15th Century and until the mid-17th Century they gradually displaced the hunter-gatherer communities of Akafula or Abatwa and established a state system that spread from the Lake Malawi region to parts of Zambia and Mozambique (Amanze 2002).

It was common practice among the Kalongas (Chewa paramount chiefs) to send out matrilineal relatives to colonise lands beyond their settlements, a practice which allowed them to spread their political influence over a wide area. Through this system they colonised the Upper Shire Valley and established themselves as 'owners of the land' (Amanze 2002). From their headquarters at present day Ulongwe in Balaka District they spread to parts of Zomba. According to the historical record it was while living along the Lake Malombe area near Ulongwe that the Chewa began to call themselves Anyanja, that is, 'people who live along the lake'.

In order to increase the land under her control, Kabuthu-Nyangu, the Chewa chief at Ulongwe, sent some of her relatives to take control of the land south, north, west and east of Ulongwe. One of those relatives, Nyani, colonised much of the Zomba District and established his headquarters at Upper Domasi in the Zomba District (Amanze 2002). The Nyanja claimed ownership of the land and its resources. It is likely that the Nyanja who settled around Lake Chilwa are an off-shoot of the Nyani chieftaincy. There is, however, no historical record of exactly when they settled the area.

As farmers, rain was critical to the livelihood of the Nyanja. Rain shrines were established where they could pray to their ancestors for rain whenever a drought threatened. The rain shrines in effect reinforced Nyanja corporate identity and its links with the Chewa are based on their common past as expressed in the myths recounted during rituals, prayers and sacrifices. It is also likely that these shrines were forums for political expression. The Chewa rain shrine in the Upper Shire area was presided over by Bimbi. Bimbi, who lived near Ulongwe, was believed to possess rain making powers and the ability to foretell calamities. These abilities were recognised by both the Nyanja and the Yao.

Bimbi shrines were frequently sought by Yao chiefs to intercede with the spirits when calamities befell communities. Thus, while the Yao chiefs assumed control of land, the Nyanja, through Bimbi, played a significant role in spiritual matters.

3.2.2 The Yao

In the 1730s and 1740s the Yao people were the greatest long-distance traders in East Central Africa, trading with the Maravi people to the southwest and with Kilwa in the northeast. Ivory was the mainstay of their trade and the chief customers were the Portuguese and Arabs (Pachai 1979). The Yao came to Malawi from Mozambique in streams of matriclans[1] between 1830 and 1860. They were driven by different motives. Mitchell (1956) indicates that the Amachinga Yao were forced by war and famine to migrate out of Mozambique to the area around Lake Chilwa and the area of Zomba town. Many of the immigrant Yao matriclans shared clan names with the Nyanja, which facilitated peaceful co-existence, as groups that share clan names regard one another as kin. As time went on, however, the Yao took advantage of internal struggles among the Nyanja to seize political power from them.

Mitchel (1959) observes that the Yao combined slave raiding and trading with agricultural production. Slave trading was essentially a dry season activity. During the rainy season they cultivated millet and sorghum. Slavery was important to the Yao because it provided a ready source of fighters for intratribal and intertribal warfare, and porterage for trade goods from the coast.

Using their experiences of long-distance trade with the East African coast, the Yao became the long arm of the Indian slave trade preying upon and occupying the land of the undefended and peaceful Nyanja of the lakeshore and Shire Highlands[2]

1 A matriclan is a group of people that claims their descent from a common ancestress whether real or fictive.

2 The Shire Highlands consist of the present-day districts of Mulanje, Thyolo, Blantyre, Chiladzulu and Phalombe.

(Alpers 1969). Historians characterise the years from 1891 and 1896 as the most disturbed period in the history of Lake Chilwa and its peoples, when the newly appointed British Protectorate administration tried to put down Yao slave trading activities by force of arms (Pachai 1979). Militarily powerful and commercially aggressive, the Yao dominated most of their neighbours for the remainder of the 19th Century.

The status and political authority of Yao chiefs and headmen was reckoned in terms of the size of their villages, which were added to by conquest as well as normal expansion (Pachai 1973). Some members of present day Yao villages are in fact descendants of slaves captured in intertribal wars and later allowed to marry Yao women. One's slave origins tend to be exposed during contests over village headship.

3.2.3 The Lomwe

The exact date when Lomwe people settled around Lake Chilwa is not known but the earliest written record dates as far back as 1859 when Livingstone is reported to have found a Lomwe village at the confluence of the Shire and Lilangwe rivers, south of Zomba Mountain. In 1861 the Universities Mission to Central Africa (UMCA) missionaries are reported to have found some Lomwe residents at Magomero, south of Zomba town. Duff MacDonald, a Scottish missionary based at Blantyre, observed Lomwe settlers along the Malawi-Mozambique area between 1882 and 1894 (Pachai 1979). By the time of the declaration of British protectorate in 1891 there were already substantial numbers of the Lomwe living in the Shire Highlands.

Several factors account for the influx of Lomwe people into the environs of the Chilwa basin. The first factor was the establishment of the tobacco estates on the outskirts of Zomba town. One such estate was Songani, founded by R.S. Hynde and R.R. Starke in 1893. This estate was taken over by the Blantyre and East African Company in 1901, which introduced curing of tobacco by fire. The first Lomwes came from Mozambique to work on these estates. Tobacco cultivation in Zomba increased rapidly and with the expansion more labour was needed, thus creating employment opportunities for Lomwe migrants from across the border. The other factor was the harsh rule imposed by the Portuguese in the Lomwe homeland between 1897 and 1907, characterised by forced labour, heavy taxation, physical brutality and plunder of life and property (Pachai 1979). The Lomwes fled to the British protectorate of Nyasaland where the treatment of Africans was perceived to be more humane. The third factor came a few years later, in the form of famine and food shortages exacerbated by representatives of the colonial administration, who plundered what little food there was. Once again, Lomwes sought refuge in the British territory. Some of these migrants entered the country through the Chilwa plain and settled in the Zomba and Chiradzulu Districts (Pachai 1979).

Some of the Lomwe who came to settle in the Lake Chilwa area were refugees from Portuguese religious intolerance. In 1894 Blantyre was devastated by a locust plague and people had no hope of harvesting anything. Three missionaries of the Church of Scotland went to the Lomwe homeland east of Lake Chilwa to purchase food. They managed to buy a lot of food in a short space of time and also saw the aftermath of the slave trade that had gone on in the area. They were also impressed by the reception accorded to them by the friendly Lomwe people (Pachai 1979). Two years later after negotiations, they opened three mission stations in the Lomwe heartland, east of Lake Chilwa. These missions, however, only functioned for six years before the Portuguese authorities ordered that they be closed in 1902. Buildings were set on fire and razed to the ground. The Protestant Scottish Mission which had a tolerant and benevolent view of colonialism was perceived as a threat to Portuguese colonialism. Since the Lomwes had in a short space of time become so attached to these missions, they perceived their destruction as an act of persecution and many more Lomwes entered Malawi and settled permanently.

The historical evidence is that the Lomwe came to Malawi in streams of small groups – not large groups – and were driven out of their homeland in Mozambique by a variety of factors. Most of Lomwe immigrants lived as guests of powerful Yao Chiefs who allocated lands to them to live on and cultivate, often according to their matrilineages. The leaders of these matrilineages were eventually allowed to exercise authority over their kin subject to the overall authority of the chief who gave them land. Over the years the three groups have intermarried so much that it may not be easy to find a 'pure' Yao, Nyanja or Lomwe.

3.3 ACCESS AND CONTROL OF RESOURCES

The main resources in the Lake Chilwa floodplains are land, water, fish and game. In pre-colonial times the basic common pool resource management system was located at the level of the chiefdom and at the level of the village. The chiefdom was in effect a federation of villages acknowledging a common ancestry. The village heads controlled access to resources. Yao village heads were known to use the granting of access to resources to build up their villages. Marital immigrants were given access to resources in the village. Simultaneously the village heads sought to retain sons of the village by granting them access rights to these resources. Not much is known about pre-colonial systems of resource management. The resources were thought to be able to sustain and renew themselves through some spiritual or divine power. Kayambazinthu and Chabwera (1999) report that God was believed to be the sole owner of the lake and what was in it, and communities under the leadership of the chiefs, who were His stewards. The fish in the lake was believed to have its own way of protecting and conserving itself. It was only when fish was scarce that they resorted to sacrifices in order for stocks to rebuild. Utilisation rather than conservation was the norm.

Binauli and Chipeta (1999), however, hint at some form of fishery management dating from pre-colonial times. The two assert that the wetlands, especially on the Kachulu side, had their own regulations on controlling fishing. There were certain seasons when fishing was prohibited until certain rituals had been performed for their ancestors. Village members were only allowed to fish using fish traps, and hook and line, but not poisonous herbs. It is not clear is who performed these rituals, whether all the different ethnic groups in the Lake Chilwa area were involved in them, and what the sanctions for non-compliance were. As the earliest settlers, it is likely that the Nyanja provided the spiritual leader, as is the case in Upper Shire valley (Amanze 2002).

3.4 POWER AND INFLUENCE OF TRADITIONAL LEADERSHIP OVER RESOURCES

The 1913 Native Administration Ordinance created a line of authority from District residents, who were British colonial officials, to villagers through village headmen and principal headman. Districts were divided into administrative sections and each section was put under the charge of a principal headman. The sections in turn were subdivided into village areas consisting of groups of villages, each under the charge of a village headman (Mitchell 1956). Under this arrangement village heads and chiefs presided over specific areas and could only grant access rights to resources within their specific areas. Similarly one could only access resources in one's village. Consequently village heads became important access points to livelihoods resources such as land, fish and game. Their control over access to these resources enhanced the status of chiefs over ordinary villagers.

At the time of independence in 1964 the new government maintained the powers of the chiefs and village heads over land and other resources subject to the overall authority of sectoral departments. The creation of irrigated rice settlement schemes in the floodplain, and the increasing opportunities that fishing and fish trading offered, brought about a new wave of immigration into the area. As the growing population increased demand for access to land and fishing rights, village heads saw an opportunity to earn an income through 'tributes', namely payments either in form of cash or fish from immigrant land seekers and fishermen to the village head for the rights to access land or fish from the lake.

Over the years, village leaders have rented out pieces of land in the wetlands with potential for rice production outside the formal government rice schemes, effectively converting a common pool into private property. Hence, three types of land holding can be identified in the Lake Chilwa floodplains:

- long-term settlers who consider themselves as owners;

- tenants of the state on government-irrigated rice schemes; and
- those who access land through the tribute system initiated by village leaders for land seekers from outside the floodplains.

The available studies (Peters 2004 and 2005, Kambewa 2005) do not reveal a pattern of conflict over control of resources, which may have been expected between the Nyanja, as the early inhabitants, and the later arrivals, that is the Yao and, Lomwe.

3.5 CONCLUSION

From my literature review, it is clear that the basin was not completely isolated from what was going on in the country before and after the colonial era. It was at a crossroad for many activities including slave trading, evangelisation, hunting and fishing. Because of the poor soils, farming was not a predominant activity in the early days. It is believed that the Nyanja were the first inhabitants of the area and these were later joined by the Yao, some of whom came as invaders and others as peaceful traders. The Lomwe who came after the Yao, were essentially refugees from the harsh treatment they were subjected to at the hands of the Portuguese in Mozambique. The Nyanja, Yao and Lomwe, who are the dominant ethnic groups in the floodplain, are culturally similar. It is likely that the similarity of the cultures of the three groups has enabled them to co-exist peacefully. There has been little contestation as to who owns or controls what resources. The existing situation is that those who settled earlier have access to and control to what they acquired when they arrived in the area. Where conflicts arise, they are essentially over boundaries.

REFERENCES

Agnew S. 1970. 'The people' pp41-44, in: Kalk M (ed) *Decline and Recovery of a Lake, Research Report 1966*–1970. University of Malawi, Lake Chilwa Co-ordinated Research Project, Government Printer, Zomba, Malawi.

Agnew S. 1972. 'Environment and history: The Malawian setting' in Pachai B (ed) *The Early History of Malawi* 28-48pp. Longman, London.

Alpers E. 1969. 'The Mutapa and Malawi political systems' in Ranger TO (ed) *Aspects of Central African History.* Heinemann, London.

Alpers EA. 1972. 'The Yao in Malawi: The importance of local research' in Pachai B (eds) *The Early History of Malawi.* Longman, London.

Amanze JN. 2002. *African Traditional Religion in Malawi: The Case of the Bimbi Cult.* Kachere Monograph No. 17. Christian Literature Association in Malawi, Blantyre, Malawi.

Binauli L and Chipeta L. 1999. *Gender, Livelihood Security Systems and Environmental Management in the Lake Chilwa Wetland and Catchment Areas.* A report done for the Lake Chilwa Wetland and Catchment Management in the Department of Environmental Affairs, Government of Malawi /DANIDA, Zomba, Malawi.

Chilivumbo A. 1970. 'Impact of development' pp49-51, in: Kalk M (ed) *Decline and Recovery of a Lake*,. Research Report 1966-1970. University of Malawi, Lake Chilwa Co-ordinated Research Project, Government Printers, Zomba, Malawi.

Kalk M. 1979. 'Focus on social problems' pp417-432, in: Kalk M, McLachlan AJ & Howard-Williams C (eds) *Lake Chilwa: Studies of Change in a Tropical Ecosystem.* Dr. W. Junk Publishers, The Hague-Boston-London.

Kambewa D. 2004. 'Patterns of access and use in wetlands: The Lake Chilwa basin.' Available from: http://www.basis.wisc.edu (Accessed on 21 May, 2012).

Kambewa D. 2005. 'Access and monopoly over wetlands in Malawi.' *Paper presented at an International Workshop on African Water Laws: Plural Legislative Frameworks for Rural Water Management in Africa, 26—28 January 2005, Johannesburg, South Africa.*

Kayambazinthu E and Chabwera E. 1999. 'The women and men of Lake Chilwa wetland and catchment areas: A gender profile.' *Report prepared for the Lake Chilwa Wetland and Catchment Management in the Department of Environmental Affairs*. Government of Malawi /DANIDA, Zomba, Malawi.

Mitchell JC. 1956. *The Yao Village: A Study in the Social Structure of a Nyasaland Tribe.* Manchester University Press, Manchester.

Mitchell JC. 1959. 'The Yao of Southern Nyasaland', in Colson E and Gluckman M (eds) *Seven Tribes of British Central Africa*, pp292-353. Manchester University Press, Manchester.

Pachai B. 1973. *Malawi: The History of the Nation.* Longman, London.

Pachai B. 1979. 'The influence of history on Lake Chilwa and its people', pp295-308, in: Kalk M, McLachlan AJ and Howard-Williams C (eds) *Lake Chilwa: Studies of Change in a Tropical Ecosystem.* Dr. W. Junk Publishers, The Hague-Boston-London.

Peters PE. 2004. 'Informal irrigation in Lake Chilwa basin: Stream-bank and wetland gardens.' Available at http://www.basis.wisc.edu.(25 May 2012)

Phiri FD. 2004. *History of Malawi: From Earliest Times to the Year 1915.* Christian Literature Association in Malawi, Limbe, Malawi.

Chapter 4

CHALLENGES IN THE MANAGEMENT OF LAKE CHILWA COMMONS

Friday NJAYA

4.1 INTRODUCTION

The importance of Lake Chilwa as a commercial fishery dates back to the 1940s, although earlier reports show trading with fish products under a barter system in the 1800s after the arrival of the Nyanja and the Yao (Kalk *et al.* 1979, Vaughan 1982). Malawi exported dried *matemba* to neighbouring Zimbabwe and Zambia in the 1980s although larger proportions targeted domestic markets (Salama and Jones 1982). The other commercially valuable fish species, *makumba* and *mlamba*, are usually destined for local markets. Schuijt (1999) estimated an annual value of US$17 million of fishing[1] from Lake Chilwa while Njaya (2002b) estimated a seine fishery[2] value of about US$8 million per year during a 'normal' fishing year. The substantial value provides livelihoods to over 9 000 households (fishers and crew). With individuals along the production chain, including fish processors and traders, the figure could be over 1.4 million people. Lake Chilwa is therefore important for sustaining the livelihoods of many Malawians.

Lake Chilwa goes through cyclic recessions due to persistent droughts that usually last three to four years (Chiotha 1996). Records indicate eight occurrences of such recessions since 1879 (Kalk *et al.* 1979, Kabwazi and Wilson 1996, Njaya 1998). When the lake recedes, the fishery collapses but recovers within three to four years after the lake refills. Four years after the 1995 recession, the estimated fish production on the lake dropped to 12 500 tons (Government of Malawi 2005).

FAO (2000) defines natural resource conflicts as disagreements and disputes that occur due to access to and control of appropriation. Conflicts related to natural resources arise from different causes (Njaya 2009). They can emerge within common property regimes between resource users and can be either violent or non-violent (Pomeroy and Rivera-Guieb 2005, Warner 2000). They usually arise because people with varying use interests and needs want to manage the resources in different ways based on their different objectives. Conflicts are natural and are avoidable. The frequency of occurrence is dependent on the demand for the resource.

---

1 This refers to value from fishing with all gear types including open (*nkacha*) and beach (*matemba*) seines, fish traps, scoop nets, cast nets and long lines.

2 Value from fishing with seines (*matemba* and *nkacha*) only.

In fishing, disagreements can also arise due to incompatible needs, or when the priorities of fishing communities are excluded during the formulation of policies, programmes and projects. Such conflicts of interest are an inevitable feature of all societies (FAO 2000).

In this chapter I will focus on fishing and conflicts on Lake Chilwa, first identifying the sources of these conflicts and then suggest ways of dealing with them more effectively than at present. My review relies heavily on research carried out between 2007 and 2009.

4.1.1 LAKE CHILWA FISHERY

Lake Chilwa is situated in the centre of the low-lying Chilwa-Phalombe plain in the southern part of Malawi (FAO 1966, Ratcliffe 1971a). The lake is the second largest in Malawi and ranked twelfth in Africa (Banda *et al.* 1997). It is shallow without any outlet and has a maximum depth of 6 m at peak water level; a reed belt of about 15 km wide surrounds the lake to the north, reduced to 1 to 2 km wide on the north-eastern side (Landes and Otte 1983).

David Livingstone, the Scottish missionary-explorer, was the first to provide information about Lake Chilwa water level changes, based on his observations. Since 1879, records show Lake Chilwa has dried up completely on six occasions, partially dried up eight times and flooded six times. They also show that fish mortalities and reduced catches occurred during the minor and moderately severe recessions of 1879, 1900, 1914-15, 1922, 1931-32, 1934, 1954, 1960-61, 1967, 1973 and 1995 (Kalk *et al.* 1979, Njaya *et al.* 1996). A 25-year record of low lake levels occurs, giving a clearer picture of periodicity of the behaviour of the lake (Agnew and Chipeta 1979).

With high temperatures of 20-39°C, significant water loss occurs through evaporation (Government of Malawi 1999). Between 1985 and 1992, evaporation averaged 1 779 mm at Domasi Rice Scheme while Makoka Agricultural Research Station recorded 1 670 mm in 1999. The two stations lie within the catchment area of Lake Chilwa. A water balance equation for the Chilwa basin shows an excessive loss of water resources through evapotranspiration processes, which makes it vulnerable to drying up during prolonged drought spells (Chavula 2000)

Humidity for the catchment area is 71% on average, with and an average of 7.1 sunshine hours and and an average wind speed of 158 km per day (Government of Malawi 1999). Its water budget depends on rainfall, water inflow from rivers and evaporation. The long-term cyclic changes of water levels of 2 to 3 m result in a partial or complete drying up of the lake (Landes and Otte 1983).

The annual rainfall for the Lake Chilwa catchment is around 1 000 to 1 200 mm per year (Nicholson 1998) and is caused by the Tropical Cyclones and Inter Tropical Convergence Zone (ITCZ), which normally occurs between November and April. The tropical cyclones generate rainfall in the catchment area as they cross the Mozambique Channel into Malawi. When the cyclone crosses the channel and moves over land, the Lake Chilwa catchment experiences widespread heavy rainfall, with accompanying flood disasters.

The 2008 Housing and Population Census shows that the catchment area has approximately 1.6 million people with a density of 321 people per km^2, an increase of approximately 50% since 1998. The catchment areas of Lake Chilwa have recorded populations of 593 167 for Zomba, 290 946 for Chiradzulu, 488 996 for Machinga and 313 227 for Phalombe (NSO 2008). The population figures indicate that the Lake Chilwa catchment is one of the most densely populated areas of Malawi.

Lake Chilwa harbours 27 fish species, but only four (the small minnow *Barbus paludinosus*, the African catfish *Clarias gariepinus*, and tilapiine species *Oreochromis chilwae* is of commercial importance (Kalk etal 1979). The valuable commercial species have high fecundity levels, reproducing at a relatively early age and can persist in the swamp, lagoon and streams as well as in the open lake. They have broad diets with considerable overlap and display opportunistic feeding behaviour (Moss 1979).

The maximum size of *B. paludinosus* (*matemba*) with high fecundity (average of 500 eggs per female) is 12 cm but they mature at 5 cm (Banda *et al.* 1997). The highest spawning peak of the *Barbus* species is from September to December (Howard-Williams *et al.* 1972). There is a fall in catches of *B. paludinosus* from September to February in the open waters probably due to spawning that occurs in rivers (Furse *et al.* 1979).

The endemic cichlid subspecies of Lake Chilwa, *Oreochromis shiranus chilwae* grows to a maximum of 15 cm (Msiska 1991). The fecundity of the cichlid is 200 eggs per female (Banda *et al.* 1997). The fish spawns in surrounding pools, springs, and lagoons and has an extended breeding season from September to May. *O. shiranus chilwae* prefers to breed in shallow waters (Furse *et al.* 1979). The fish species have two main spawning periods in an extended breeding that are suitable for the fluctuating environment. *O. shiranus chilwae* is more associated with the open water than *Barbus* and *Clarias* species, which move into the swamps to breed. Moss (1979) observed that *O. shiranus chilwae* has relatively lower resistance as shown in the delayed recovery of its populations after recessions.

The Lake Chilwa catfish *Clarias gariepinus* (*mlamba*) is the only fish species that is smaller than other catfishes (Banda *et al.* 1997). Fecundity of *C. gariepinus* is high with 180 000 eggs per female for each breeding season. The unsexed *C. gariepinus* in Lake Chilwa matures earlier at 18.5 cm than the female ones from other lakes. The female *C. gariepinus* from Lake Kariba, Lake Tana and reservoirs in Burkina Faso mature at 34, 30.5 and 37.5 cm, respectively (Teugels 1986). Spawning of *C. gariepinus* usually begins in September and continues during the rainy season in flooded delta areas and they return to the river or the lake soon afterwards (Furse *et al.* 1979).

The fishery is predominantly artisanal[3]. Commercial fishing operations using trawlers took place in the 1970s and 1980s but these fishing operations failed mainly due to highly unstable lake levels, poor infrastructure and lack of effective marketing strategies. Since the 1995 recession, about 6 000 operators on average are engaged in fishing activities. Schuijt (1999) estimated that Lake Chilwa had 3 720 fish traders, 17 boat builders and a considerable number of people involved in other ancillary industries.

Lake Chilwa fishers use fish traps, gillnets, seine nets and long lines. Fish traps are the most common because they are less expensive as they are made of local materials compared to other gear types that need synthetic materials. Gillnets are also cheaper than seines. Two types of seines exist in Lake Chilwa: *matemba* seines that are adapted to the ecological conditions of the lake, and *nkacha* seines that the Lake Malombe in-migrants operate in the lake. Fishers use fish traps and seines mainly to catch *matemba* while they use gillnets and long lines to target large fish species such as *makumba* and *mlamba*.

Between 1976 and 2010, the annual fish production of Lake Chilwa was 12 000 tons on average with production exceeding 25 000 tons in 1979 and 1990 (Njaya 2009). During the recession period (1995-1996), fishing operations were suspended or reduced to a lower scale in river mouths. Contribution of fish landings from the lake to the national production levels varies on an annual basis with an average of 15% from 1993 to 2010 (excluding 1995 and 1996 when the lake dried up). After the previous recession, the highest contribution of fish supply was recorded in 1999 at 27%.

4.2 FISHERIES MANAGEMENT SYSTEMS

4.2.1 Management regime changes on Lake Chilwa

A number of management regimes have over the years been tried in efforts to protect Lake Chilwa fisheries from overexploitation, including traditional fisheries management, centralised fisheries management and co-management. The changes occurred mainly due to the influence of colonialism and the economic shift from a subsistence to commercial economy in the 1970s.

Little is known about how fisheries on Lake Chilwa were managed before the colonial rule. It is possible that open access was the norm since fish were regarded as naturally given to the people for their utilisation through powers vested in their

[3] In this chapter, we use the terms 'artisanal' or 'small-scale' fishery interchangeably by referring to the nature of the fishing gears and types of craft used. The definition may not apply to other lakes in Malawi or elsewhere. Fish traps, gillnet and long lines are categorised as artisanal while *matemba* seine nets, *nkacha* and trawl nets are commercial. Dugout canoes are regarded as artisanal while planked boats are commercial.

traditional leaders (Dissi and Njaya 1995, Njaya 2007). Regulatory mechanisms may not have been necessary since population levels were too low for the fisheries to be over exploited.

Following the introduction of British Indirect Rule and the Native Authority treasuries in 1933, chiefs in lakeshores districts were empowered to regulate access to the lake and they could limit access to the fishing grounds. It is, however, debatable whether the revenue collected was used for managing fisheries resources or treated as just another form of revenue for government. Chirwa (1996) has asserted that the chiefs could levy taxes on canoes and impose quotas on trees to be cut for canoes; although this assertion is based on the research conducted on Lake Malawi, it could arguably also apply to Lake Chilwa.

Centralised fisheries management (1946-1995) was introduced by the colonial government to ensure sustainability, equity and efficiency in the management of the fisheries resources. Several regulations were formulated but enforcement was a problem. Some of the key regulations that governed exploitation of the fisheries resources on Lake Chilwa were:

- Minimum size of tilapia (*makumb*a) was set at 100 mm to allow selling of fish of larger sizes that have bred at least once.
- Maximum headline length for *matemba* seines was set at 300 m to regulate size of the seined area.
- Minimum mesh size for gill nets was set at 69 mm to ensure catching of *makumba* larger than 100 mm.
- Licensing of *matemb*a seines and gill nets was introduced for controlled access.

Although the government introduced licensing of beach seines and gillnets on Lake Chilwa in the 1960s the exploitation of the fishery resources remained largely unregulated (Ratcliffe 1971b). The main problem was that the licensing system was considered a revenue generation scheme and not as a management tool. The capacity of the Department of Fisheries to enforce fisheries regulations was also hampered by budgetary constraints.

The Lake Chilwa co-management scheme was a response to a lake recession after a three-year drought period in 1992-1994. As was the case in 1968 when a similar recession had occurred, a strategy was necessary to facilitate recovery of the collapsed fishery. The first proposed strategy involved conducting artificial restocking by breeding *O. shiranus chilwae* in ponds at Domasi and then releasing the fish into the lake. However, this could be expensive in terms of required resources such as funds, human and technical skills. Furthermore, it was practically difficult to breed *matemba* (*B. paludinosus*) in ponds for restocking in the lake, hence natural restocking during the 1995 recession was the only suitable strategy.

Traditional leaders in partnership with the Department of Fisheries suggested that for a successful natural restocking programme, there was a need to conserve all fish stocks in Mpoto lagoon located to the south-east of Lake Chilwa, and reservoirs along the influent rivers that include the Domasi, Likangala, Thondwe,

Phalombe, and Sombani in Malawi, and Mnembo and Chimazi in Mozambique. The aim was to have the conserved fish stocks repopulate the lake after refilling. The fishers had a similar idea, as that was what they had been doing during past recessions. A similar programme had been adopted in 1968.

In this recovery management strategy, collective action with the involvement of the river-based fishing communities was important, as this would make it easier to enforce regulations. The local leaders and the Department of Fisheries organised meetings in several villages located along the major influent rivers seeking to work out an effective partnership for enforcement of rules, which they formulated and then publicised on radio and newspapers:

- ban on the use of poisonous plants (*katupe*) for fishing in rivers flowing into Lake Chilwa;
- prohibiting seining operations in all influent rivers and the lagoon; and
- restricting use of seines from 1996 to 1997.

Upon refilling of the lake in 1996, the Department of Fisheries seized the opportunity to introduce and expand the co-management framework to all the lake-based fishing communities. The Department organised a meeting at Chilema in 1996 with the lake-based traditional chiefs, at which they reviewed the rules and regulations by introducing closed season for Lake Chilwa from December to April, banning *nkacha* seines, enforcing licensing of seines and gillnets, and enforcing the use of operating gillnets with meshes over 50 mm.

Towards the end of 1996, the Department of Fisheries sensitised local households on the need to form Beach Village Committees (BVCs), which recruited fishers, mostly those operating fish traps, gillnet and long line. A total of 48 BVCs were established. In the following year, 1997, six chiefs and four group village heads around the lake formed the Lake Chilwa Fisheries Management Association. However, the association had no elected fisher to represent the interests of fishing households and was criticised for lack of representation (Lowore and Lowore 1999, Njaya 2002a). This criticism implied that the reviewed rules were not in the interests of the fishers but for the benefit of the association, especially for charging penalties in the form of fines imposed on illegal fishers. Additionally, the exclusion of seine fishers in the BVCs meant that there was limited participation of other user groups in the co-management arrangement. However, through meetings conducted during a programme funded by the United States International Development Agency (USAID), many local seine fishers joined the co-management programme as partners through their own association and did not want to recognise the existing Lake Chilwa Fisheries Management Association composed of the chiefs.

4.3 FISHERIES-RELATED CONFLICTS IN LAKE CHILWA

The Food and Agriculture Organisation (2000) defines natural resource conflicts as disagreements and disputes that occur due to access to and control of appropriation. Conflicts can emerge within or between resource users and can be either violent or non-violent (Pomeroy and Rivera-Guieb 2005, Warner 2000). Usually the conflicts arise because people have varying utilisation of the commons such as fisheries or want to manage them in various ways based on their different objectives. Singh (2002) observes that conflicts and disturbances arise when there are varied interests of people in a particular resource. Generally, conflicts can occur under any managerial regime as it is impossible to have homogeneous interests among groups of people. In the Lake Chilwa basin conflicts have revolved around cultural issues, the environment, use of technology and the application of government policy.

Conflicts mainly revolve around the use of fishing technology or are associated with policy and legislative frameworks.

4.3.1 Conflicts over use of fishing technology

Conflicts between small-scale[4] and large-scale fishers have been prevalent on Lake Chilwa in the recent past. The most prominent conflicts arise between gillnet fishers and *nkacha* (open water seine) net operators. The Yao from Lake Malombe are associated with the introduction of *nkacha* seines on a number of lakes, including lakes Chilwa, Chiuta, Malombe and Malawi. Gillnets on the other hand have been long part of fishing tradition on Lake Chilwa. The recurrent complaint heard from gillnet fishers during fieldwork in 2009 was that *nkacha* seine fishers damaged their seines in open waters. In the words of one gillnet fisher (Njaya 2009):

> The problem with seine fishers is that they operate their seines everywhere including in places where we set our gill nets. In the past, gillnet fishers set their gears in the open waters since seine fishers could not access such fishing areas. However, nowadays with the coming of the *nkacha* fishers from Lake Malombe the whole lake is under exploitation. The problem is that the seining operations in the open waters damage our gill nets and it becomes difficult to identify who is responsible for such damage. We, the gillnet operators, end up buying new gillnet materials on a higher frequency, which is costly.

Most chiefs support the in-migrants who own *nkacha* seines for socio-economic reasons. The *nkacha* fishers catch larger amounts of fish for sale to fish mongers and also offer certain portions to the chiefs as *cha-kwa-mfumu*[5]. The presence of the *nkacha* fishers is furthermore associated with increased economic activities,

4 In the context of Lake Chilwa, the small-scale fishers refer to those using fish traps, cast nets, scoop nets and long lines while the large-scale fishers operate seine nets and gillnets.

5 A portion of fish catch given to a local leader as a way of tribute. In other places like Mangochi they call it *mawe* (Hara *et al.* 2002) while on Lake Chilwa it is known as *cha-kwa-mfumu*.

such as shops, rest houses, restaurants and transport, which create local employment for local young men. Demand also increases for fishing related activities such as boat building and boat engine repairs. As one chief explained in an interview in 1998 (Njaya 2009):

> I feel it is important that we allow the migrant fishers to operate in our areas. Apart from their fish landings, which are larger due to use of *nkacha* seines, they also use land, which could be lying idle. As you know, many of these seine fishers from Lake Malombe are rich and hence have adequate capital to invest in our villages mainly in terms of fishing and farming in which case they employ young men, which I believe is necessary.

However, the resident fishers argued that the chiefs support the in-migrant fishers because the chief benefit from the *cha-kwa-mfumu* system. Although an element of jealousy cannot be ruled out in these assertions, there is no doubt that *nkacha* has been reported to be an ecologically destructive gear by researchers on Lake Malombe (Banda *et al.* 1997).

Access to distant fishing grounds involves use of bamboo poles (*miponda*) and paddles (*malemu* or *nkhafi*). The migrant fishers introduced the use of paddles in the 1980s. Traditionally Lake Chilwa fishers operate seines in inshore waters or localised areas where they construct temporary structures for dwelling (*zimbowera*). The migrant fishers claim offshore fishing targets of breeding and immature *matemba* while traditional fishers catch the mature ones migrating to shallow and marshy areas. As one old seine fisher at Khanda stated (Njaya 2009):

> The *matemba* breeding stock and immature fish are located in deeper waters after which they migrate to shallow and marshy areas for feeding and hiding. These are the areas I target for seining but you see that our colleagues from Lake Malombe go to the deeper areas using their plank boats propelled manually by use of *malemu* [paddles] instead of *miponda* [bamboo poles] where I believe are spiritually reserved sites [sanctuaries] for fish breeding and yet they operate their seines there. The migrant *nkacha* fishers always fish in the sanctuaries. This is unfair to our tradition and culture, which involves giving respect to our spirits. The spiritually reserved sites may or may not actually exist; but their perceived destruction at the hands of foreigners reinforces the argument against migrant fishers.

While some scholars believe Lake Chilwa cannot be overfished due to the resilient nature of the ecosystem (Kalk *et al.* 1979, Sarch and Allison 2000), the local fishers strongly believe that overfishing can occur on the lake, as a certain woman seine owner at Mchenga stated (Njaya 2009):

> When I started fishing in 1997 with the seine net, which my father gave me, I used to land high catches with subsequent high daily revenue from fish sales. I bought a car and built a house within one year. However, the situation changed from 2000 when the daily sales were declining to the point of depending on income from the house rent and car hire services.

These divergent views on the possibility of overfishing of Lake Chilwa fish stocks contribute to conflicts between the migrant seine fishers and the resident fishers using traditional seines and other passive gear types (fish traps, long lines and gill nets).

4.3.2 Conflicts associated with policy and legislative frameworks

National policy and legislative frameworks governing natural resources are frequently in conflict with each other. In the Lake Chilwa Basin, for instance, ministries responsible for agriculture and irrigation promote cultivation along riverbanks for food security, which has the potential of causing siltation of the rivers and the lake due to soil erosion, with negative impacts on the spawning grounds of fish. The cultivation of river banks actually violates the National Environment Policy of 1996.

Whereas the Water Resources Policy of 1994 aims at ensuring that all citizens of Malawi have convenient access to water of sufficient quantity and quality, water extraction may, if not properly managed, affect the water level of rivers which flow into the lake, thereby affecting aquatic life such as fish.

Legal pluralism with conflicting elements seems to apply to the Lake Chilwa ecosystem and yet not too long ago local fishing-farming households used to manage the ecosystem in a holistic manner, as a group of fishers on Chisi Island stated (Njaya 2009):

> We used to utilise and manage Lake Chilwa as one ecosystem with fish, water, birds, grass, trees and land with guidance from our traditional leaders unlike nowadays when I hear about BVCs for fish, *Mwayi wa Mbalame* Association for birds, village natural resource committees for trees and so on. During drought, we offered sacrifices to have water, which refilled the lake and subsequently have fish and grass...

The above statement shows how difficult it may be to introduce policy frameworks to guide management of fish, water, wildlife (mainly birds), land, and forests. It also demonstrates the level of difficulty to have a co-management arrangement for each one of the natural resources. Further conflict may be triggered by increased population growth and the introduction of new fishing technologies. Politics may hinder co-management, as in some areas the same people get elected into key positions in the community-based natural resource committees. Eventually there is limited efficiency in resource utilisation and effectiveness in achieving the intended objectives of co-management arrangements.

4.4 CONCLUSION

In this review, I have revealed that fisheries management of Lake Chilwa revolves around addressing conflicts which may relate to operational and socio-economic parameters and policy parameters. There is no consensus among the fishers regarding sustainable economic utilisation of Lake Chilwa. On the one hand, migrant fishers are blamed for overfishing the lake through their use of the open water *nkacha* seine nets, while on the other hand fish traders, who are mostly women, benefit from the larger catches by the *nkacha* fishers. Consequently, larger fish landings and volume of sales make significant contributions to the local economy.

The issue about some traditional leaders getting regular fish portions or money (a 'tribute') for use of their beaches by the migrants seine fishers remains difficult to handle. Traditionally, this practice is acceptable despite challenges in determining to what extent it constitutes corruption. Being one of the key sources of conflicts between traditional leaders and BVCs in the management of the fisheries resources, a policy recommendation is needed that would recognise the traditional practice of giving out fish portions to traditional leaders without compromising sound resource management strategies.

I recommend a fishery strategy that takes into account various fishing practices and ecological concerns for an improved and sustained livelihood of the vulnerable fishing communities in the Lake Chilwa basin. In line with this recommendation, it is necessary to formulate environmental management policies that would cover the whole catchment area of the shallow ecosystem of Lake Chilwa (UNEP 1981). This approach emphasises the need to manage the whole lake, wetland and its catchment area as one ecosystem, as the lakeshore based communities used to do in the past when they considered management of all the resources (land, water, birds, fish, wildlife and forestry resources) without looking at a particular sector. However, consideration of the pressure from increased human population and more effective seining technology is necessary. In addition, I recommend the formulation of an integrated management plan that aims to reduce deforestation in the watershed areas and reduce pollution into the lake, and to prohibit fishing methods that involve the removal of aquatic vegetation.

REFERENCES

Agnew S and Chipeta C. 1979. 'Fish and fish trading' pp345-368 in Kalk M, McLachlan AJ, Howard-Williams C (eds) *Lake Chilwa: Studies of Change in a Tropical Ecosystem*:. Dr. W. Junk Publishers, The Hague-Boston-London.

Banda M, Bulirani A, Kachinjika O and Njaya FJ. 1997. 'Fisheries of Malawi', in Rashidi BBR (ed) *The Fisheries Resource Management and Extension Handbook: a Field Manual for Use by Fisheries Extensionists and Others Engaged in Fisheries Resource Management*: 47-57. Department of Fisheries, Lilongwe.

Chavula GMS. 2000. *The Evaluation of the Present and Potential Water Resource Management for the Lake Chilwa Basin.* Land Tenure Centre, University of Wisconsin United States.

Chiotha S. 1996. 'Development of management plan for Lake Chilwa' pp5-7, in Njaya F, Kabwazi H, Chiotha S (eds) *Management Plan for Lake Chilwa and its Catchment Area.* Government of Malawi /MAGFAD/University of Malawi, Zomba, Malawi.

Chirwa WC. 1996. 'Fishing rights, ecology and conservation along southern Lake Malawi: 1920-1964', *African Affair* 95: 351-377.

Dissi C and Njaya FJ. 1995. 'Development of community-based fisheries management programme for Lake Chiuta.' *Paper presented at the regional co-management workshop held at Kariba, Zimbabwe, 24—27 August 1995.*

Food and Agriculture Organisation (FAO). 1966. *Report to the Government of Malawi: A Programme for Fisheries Development Based on the Work of HLF Renson.* UNDP/TA Fisheries Development Adviser. FAO No. TA2239. United Nations Development Programme, FAO, Rome.

FAO. 2000. *Conflict and Natural Resource Management.* FAO, Rome. Available at:http: //www.fao.org/forestry/foris/pdf/conflict/conf-e.pdf (Accessed on 25 May 2011).

Furse MT, Kirk RC, Morgan, PR and Tweddle D. 1979. 'Fishes: Distribution and biology in relation to changes' pp175-208, in: Kalk M, McLachlan AJ and Howard-William C (eds) *Lake Chilwa: Studies of Change in a Tropical Ecosystem*:. Dr. W. Junk Publishers, The Hague-Boston-London.

Government of Malawi. 1999. *State of the Environment Report.* Lake Chilwa Wetland and Catchment Project, DANIDA, Zomba, Malawi.

Government of Malawi. 2000. *Annual Fish Catch Records for Lake Chilwa.* Department of Fisheries, Ministry of Natural Resources and Environmental Affairs, Lilongwe, Malawi.

Government of Malawi. 2005. *Annual Fish Catch Records for Lake Chilwa.* Department of Fisheries Ministry of Natural Resources and Environmental Affairs, Lilongwe, Malawi.

Howard-Williams G, Furse M, Schulten-Senden C, Bourn D and Lenton G. 1972. *Lake Chilwa: Studies on the Tropical Freshwater Ecosystems.* IBP/UNESCO. Available at: http://www.fishbase.org/References(15 May 2012)

Jamu DM and Brummet RR. 1999. 'Fish reproduction in Lake Chilwa with special emphasis on *Barbus paludinosus* and the status of the watershed.' *State of the Environment Study No. 5.* Lake Chilwa Wetland Project, Zomba, Malawi.

Kabwazi HH and Wilson JG. 1996. 'The Lake Chilwa environment' pp 91-108, in: van Zegeren K and Munyenyembe MP (eds) *A Report of the 1996 Ramsar Site Study,.* Assemblies of God Literature Press, Zomba, Malawi.

Kalk M, McLachlan AJ and Howard-Williams C. 1979. 'Lake Chilwa: Studies of change in a tropical ecosystem', *Monographae Biologicae* 35: 17-227.

Landes A and Otte G. 1983. 'Fisheries development in the Lake Chilwa and Lake Chiuta region in Malawi.' *Report prepared for the Germany Agency for Technical Cooperation, Stufyfond PN 85.2218.7-01.100.* MAGFAD, Zomba, Malawi

Lowore J and Lowore J. 1999. *Community Management of Natural Resources in Malawi.* Government of Malawi/DANIDA, Zomba, Malawi.

Moss, B. 1979. 'The Lake Chilwa Ecosystem – A limnological overview' pp 399-416, in: Kalk, M., McLachlan AJ and Howard-Williams C (eds) *Lake Chilwa: Studies of Change in a Tropical Ecosystem.* Dr. W. Junk, The Netherlands.

Msiska OV. 1991. 'A synoptic review of the biology and culture of Oreochromis shiranus shiranus and Oreochromis shiranus chilwae', Malawi Government Available at: http://www.fishbase.org. (Accessed on 20 December 2012).

Nicholson SE. 1998. 'Fluctuations of rift valley Lakes Malawi and Chilwa during historical times: A synthesis of geological, archaeological and historical information', p208, in: Lahman JT (ed) *Environmental change and response in East African Lakes.* Kluwer Academic Publishers, The Netherlands.

Njaya FJ, Chiotha S and Kabwazi H. 1996. 'Lake Chilwa Management Plan.' *Proceedings of a workshop on the development of Lake Chilwa management plan, Zomba, Malawi.* DANIDA Project, Malawi.

Njaya FJ. 1998. 'Co-management programmes in Malawi: A comparative case study for Lakes Malombe, Chilwa and Chiuta' pp. 260-269, in: Eide A and Vassdal T (eds) *The Ninth Biennial Conference of the International Institute of Fisheries Economics and Trade.* Norwegian College of Fishery Science, Norway.

Njaya FJ. 2002a. 'Fisheries co-management in Malawi: Implementation arrangements for Lakes Malombe, Chilwa and Chiuta' pp. 9-25, in: K. Geheb and MT Sarch (eds) *Africa's Inland Fisheries: The management challenge*: 9-25. Fountain Publishers, Uganda.

Njaya FJ. 2002b. 'Review of management measures for Lake Chilwa.' *Project paper submitted to the United Nations University (UNU).* UNU, Reykjavik, Iceland.

Njaya FJ. 2007. 'Governance challenges for the implementation of fisheries co-management: Experiences from Malawi', *International Journal of the Commons* 1(1): 123-139.

Njaya FJ. 2009. 'The Lake Chilwa fishing household strategies in response to water level changes: migration, conflicts and co-management.' *PhD thesis.* University of the Western Cape, South Africa.

National Statistical Office (NSO). 2008. *Malawi Population and Housing Census.* NSO, Zomba.

Pomeroy RS and Rivera-Guieb R. 2005. *Fishery Co-management: A Practical Handbook.* International Development Research Centre, Canada.

Ratcliffe C. 1971a. 'Experimental small craft pair trawling: Lake Chilwa 1970-71. Ministry of Agriculture and Natural Resources, Extension Aids Section', *Fisheries Bulletin* 1: 27-31. Ministry of Agriculture, Government of Malawi, Zomba, Malawi.

Ratcliffe C. 1971b. 'The fishery of the Lower Shire area, Malawi', *Fisheries Bulletin* 3: 21-21. Ministry of Agriculture, Government of Malawi, Zomba, Malawi.

Salama RD and Jones DR. 1982. 'Fish exports from Malawi – A supply study.' *Malawi Export Promotion Council desk research study report.* Zomba, Malawi.

Sarch M and Allison EH. 2000. 'Fluctuating fisheries in Africa's inland waters: Well adapted livelihoods, maladapted management.' *Paper presented at the Tenth International Fisheries Economics and Trade conference, Oregon State, 10-14 July 2000.*

Schuijt KD. 1999. 'Economic valuation of the Lake Chilwa wetland.' *State of the Environment Study No. 11, Lake Chilwa Wetland Project.* Lake Chilwa Wetland Project, Zomba, Malawi.

Singh SK. 2002. 'Conflicts and disturbance: A reason to change: Lessons from community-based natural resource management institutions in Orissa, India. *Paper presented at the 9th Biennial Conference of the International Association for the Study of the Common Property, held at Victoria falls, Zimbabwe, 17—–21 June 2002.*

Teugels GG. 1986. 'A systematic revision of the African species of the genus *Clarias* (Pisces; Clariidae)', *Ann. Sci. Zool* 247: 15-48.

United Nations Environment Programme (UNEP). 1981. 'The ecology and utilisation of African waters', *Reports and Proceedings Series1:112.* UNEP, Nairobi.

Vaughan M. 1982. 'Food production and family labour in southern Malawi: the Shire

Highlands and Upper Shire Valley in the early colonial period', *The Journal of African History,* 23(3): 351-364.

Warner M. 2000. 'Conflict management in community-based natural resource projects: experiences from Fiji and Papua New Guinea', *Working paper, 135*, Overseas Development Institute, United Kingdom.

Chapter 5

TENURE SECURITY AND LAND USE IN THE LAKE CHILWA WETLANDS

Daimon KAMBEWA

5.1 INTRODUCTION

This chapter is based on a study that was conducted in the Lake Chilwa wetland in Malawi's southern region from 2003 to 2006 to understand how access and use of land in the wetland is shaped. The study was part of a larger research project in the Lake Chilwa Basin funded by USAID Washington through the BASIS[1] Project. Other parts of the study examined the formal irrigation systems of Mpheta and Likangala schemes in the wetland, and informal irrigation along the Likangala River. I used a combination of quantitative and qualitative methods, which included a survey, participant observation and individual interviews, to collect primary data on rights of use and patterns of access to land and water.

While wetlands provide many uses, communities in the Lake Chilwa wetland mostly value agriculture.Fishing is an important activity, but the owners of fishing gear are often urban-based entrepreneurs employing local labour. Agriculture takes place as formal and informal irrigation. Informal irrigation is locally called *dimba* farming and is carried out especially in the dry season on customary land. Formal irrigation encompasses government schemes that were established from the 1960s to 1970s and self-help schemes that were constructed in the 1980s. Both formal and informal irrigation are promoted as a strategy for rural development to improve rural income and food availability. Recently there has been a policy shift away from government-owned (formal schemes) to farmer-managed schemes and informal irrigation because government cannot sustain formal irrigation schemes (Government of Malawi 2001). It is believed that wetland cultivation can transform rural livelihoods if people receive relevant advice, technology and input support.

Farmers who have access to land in wetlands have stable cash flow and they are food secure because the nature of agricultural production makes it possible to generate cash and harvest food throughout the year. The farmers are able to produce crops such as maize, rice, water melons, vegetables, tomatoes and sugarcanes under both rain-fed conditions and irrigation.

[1] Broadening Access and Strengthening Input Market Systems Collaborative Research Support Project on *Institutional Dimensions of Water Policy Reform in Malawi: Addressing Critical Land Water Intersections in Broadening Access to Key Factors of Production.*

Understanding access to and use of land and water in wetlands is essential for appreciating the level of tenure security. The perception that persisted through the colonial and post-independence era in Malawi's policy making circles is that customary land tenure does not provide adequate security for the user, resulting in under investment and reliance on primitive technology (Government of Malawi 2002). In the wetland secure tenure is important for members of the community to co-operate and commit themselves to the sustainable use and management of resources.

5.2 EVOLUTION OF LAND TENURE IN THE LAKE CHILWA BASIN

The Lake Chilwa basin is situated in the southern region of Malawi. The people are mostly Nyanja, Yao and Lomwe. They practise matrilineal inheritance and succession and matrilocal residence, which occurs when the husband moves to his wife's village. A group of sisters forms the core of a village. Usually the eldest brother becomes the village head. Men who marry into the village become members by virtue of marriage and for the duration of the marriage. Primary land rights belong to the women.

There is no evidence of scholarly research into pre-colonial tenure regimes in the wetlands. It is likely that the wetland was an open-access common pool resource to which people belonging to villages in the surrounding areas had access. No individual or family could claim ownership of it or its resources. The wetlands existed for the common benefit of the people who lived around it as sources of water, animals, and birds. The small human population of that period might not have posed a threat to the resources like they do now.

According to Phiri (1984) changes in land tenure in the Chilwa basin could be traced back to the invasion of the Ngoni in the late 1870s. As a result of the invasion, the Nyanja and Yao deserted their land and fled into the mountains of Zomba and Machinga. The deserted land effectively became 'no man's land'. Between 1881 and 1894 much of the land was acquired on behalf of European settlers by the British Consul, John Buchanan. These land acquisitions were confirmed with 'certificates of claim' by the first British Commissioner and Consul General, Sir Harry Johnston. The colonial regime regarded these certificates as conferring freehold status to the owner of the certificate. By 1890, most of the land was converted into a number of freehold estates called the Shire Highlands estates. Among them was the single largest block of 169 000 acres, obtained by Buchanan for Alexander Low Bruce, the son-in-law of David Livingstone, the Scottish missionary and explorer (Peters and Kambewa 2007). The colonial era introduced to the country the alien Western notion of private property rights, a notion that was in conflict with the traditional communal view of land as belonging to a group or community.

With regard to the Lake Chilwa wetlands, communities held land under the authority of the chiefs and the land was managed according to the local traditions and customs (Vaughan 1978). In the 1960s more wetlands were converted from customary to public to create irrigation schemes. Later, in the 1980s, in response to increased demand for land in the wetlands in the 1980s, chiefs partitioned the land outside irrigation schemes and treated it as belonging to specific lineages and families (Kambewa 2006). The lands that were allocated in this way were treated as *de facto* private lands. The chiefs were in effect behaving as landlords. The demand for land coincided with the decline in livestock population due to diseases in the 1970s and 1980s. The situation was exacerbated by high levels of cattle theft in the period immediately following the return to multiparty democracy in 1994. The demand for land also intensified as more people were attracted to the benefits from irrigation schemes, which had been established in the area in the 1960s. Women were allowed to acquire land in the schemes. This practice was in line with the existing tradition in the dry land fields where women inherit the land from their mothers and grandmothers. The current property rights regime in the Lake Chilwa basin is one in which customary land holding coexists with private property rights held as leasehold or freehold.

5.3 TENURE SECURITY IN THE LAKE CHILWA WETLANDS

Access to land in the wetlands is deeply embedded in social and power relations, with the chiefs and their families as the major 'institutions' through which people negotiate access (Kambewa 2006). The main modes of access are inheritance and permission from the chiefs. The dominant mode, however, is inheritance within the lineage. Women have primary rights to land. Under the uxorilocal principle men access land in the matrimonial village through their wives. In this context the land tenure system discriminates against men. Another example on how social relations shape access is the practice of lending and borrowing land, which occurs mostly among core lineage members, close friends and those related by marriage. It is known that some individuals increased their chances of access by marrying into the family of the chief. While some members loaned out gardens to help fellow members of the lineage to grow crops in the dry season, it was noted that this was done primarily to keep the gardens clean for rainy season planting. Access is rooted in reciprocity, where people feel they should be helped because they also help others (Mtika 2001). However, some landholders did not want to loan out land because they feared that borrowers would not return the gardens, implying that access through borrowing is tied to the trust that exists between people.

While social ties are the primary conduits for access, the second mode, which is allocation by chiefs, implies that power shapes access and use of wetlands. The chiefs allow access to land in the wetlands to non-core members of any lineage or new migrants, locally called *obwera*. As it will be explained in the sections below, these farmers only have secure access if they pay an annual rent to the relevant chief. Thus, being a resident of a village does not guarantee access; rather, access depends on one's willingness to serve the interests of those in power.

The explanation of how chiefs appropriated common land as private property lies in their ability to use their traditional powers to accumulate land to themselves by suppressing the interests of those with whom they had weak relationships, for example, the new migrants and those members of the village who did not belong to the core lineage. A case in point is related by Kambewa (2006), where a chief has been seizing land since the 1980s from immigrants who came into the area to work in irrigation schemes or to fish in the lake. These immigrants were known as *obwera* despite having lived in the area for more than 40 years. The chief appropriated the land from the 'strangers' by claiming that the land should be given to primary settlers. The 'strangers' withdrew but the primary settlers did not take up the land because they feared that they would be bewitched by the 'strangers'. Failure by the primary settlers to take the land gave a chance to the chief to appropriate the land for himself. The chief announced that in-migrants were free to cultivate in the wetland again, but under the conditions that they would become registered tax-payers in the village and that they would meet some obligation to him as the 'owner' of the land. Many did so, migrating from the surrounding villages and far places such Mulanje, Phalombe, Balaka, Ntchisi, Nsanje Districts and even from Mozambique.

The main obligation to the chief of the land seekers was payment of tribute in return for the land allocated. The tribute was in the form of bags of rice from the rainy-season harvest. It was locally called *chothokoza* (literally meaning 'thanks'). Committees were responsible for registering people, showing them plots and collecting bags of rice after harvesting. Some farmers paid one 50 kg bag of rice per garden, while others paid one bag regardless of the number of gardens. The behaviour by the chiefs is often referred to as 'rent-seeking' by political scientists and economists, where the chiefs behave like landlords, instead of being trustees of the land.

The rising demand for the wetlands, especially in view of recurrent droughts and dwindling dry-land fields, may well increase the tendency of these chiefs to appropriate more wetlands. If chiefs succeed in consolidating land for themselves, one may expect increased land insecurity among the villagers, especially new immigrants. Already, land concentration in the wetlands has generated social conflict, social divisions, and splits within villages, including arguments over who should be the rightful chief or village head. Arguments in the past about who should succeed to leadership of a particular village have revolved around whether the candidate's ancestors were primary settlers or not. This implies that these arguments were not about chieftaincy *per se* but about who should have authority over the valuable wetlands.

The practice of giving part of a harvest to a chief has been more closely associated with dry-land fields than wetlands (Government of Malawi 2001). One chief claimed he was correct to take tribute as compensation for the services he rendered such as settling land and family disputes and organising agricultural inputs. The chief claimed that members of the core lineage used descent and kinship to him to claim these services, but the immigrants had to pay in the form of tribute. Another chief justified the practice as the only way he could get food and income since he could not work his land, as he was always busy performing his duties as chief. In 2002/03, this chief reported that he received 40 bags of rice and sold them for the equivalent of US$5 each. He used the money to buy iron sheets for his two houses. Another chief received 26 bags through tribute from people. The chiefs defended the tribute practice saying they collected bags of rice to sell and use the money for development work that benefited the whole community, for example, construction of bridges and paths. This was, however, disputed by most respondents, who said that the chief always asked his people to make financial contributions but the community still had to construct the bridges. The outcry from the people was that they were being exploited. In fact, the people complained that they were required to pay cash in order to be shown the land for gardens and to construct their huts. This once-off 'fee' varied from US$1 to US$10.

The tribute was an annual obligatory rent for land in wetlands. At the time of my research, some village heads who had not already adopted the practice, were considering adopting it on the grounds that they did not receive adequate compensation from working as village heads. They were also attracted by the considerable rental income earned by their peers. What was traditionally viewed as position of service to the community is increasingly being viewed as a job for which one must be paid adequate compensation. The tribute was not uniformly applicable: fellow village heads, the elderly, or core members of the lineage and group village heads, were exempted from it. The practice is widely contested. One village head said that he stopped the practice because some of his people abused it and people in the area showed displeasure. The practice was indeed prone to abuse as noted from reports that committees collected rentals for the gardens in addition to bags of rice. They also collected more bags of rice than they passed on to the chief. For example, one of the chiefs indicated that in the year 2003 he got 26 bags while the committee said they collected 46 bags, and yet there were over 300 households in the area who paid several bags each.

The tribute practice appears to be an old practice in the Lake Chilwa wetlands as it was similar to an old custom, locally known as *kuphikira mowa*, in which new families or households settling in a village prepared beer and other foods to thank the chief for giving them the land to settle and cultivate. In the colonial period, a family or household would move out of its village seeking new land to settle and cultivate in another village. The chief in the new village gave land without demanding any payment. During the harvest period, the new family prepared beer and foods and called the chief and his subjects to a feast. At the end of the feast, the family organised gifts for the chief, which could be livestock or crops. *Kuphikira mowa* marked a time when a family or household was officially accepted as members of the village. In the subsequent years, every household gave produce to the chief as thanks for the land they cultivated. In some cases chiefs were thanked for the rain-calling practice, where they were associated with powers to call rain. *Kuphikira mowa* does not take place any more but chiefs still receive gifts every year.

Kuphikira mowa could have formed the basis of the tenancy labour system that came to be known as *thangata*. According to Mkandawire (1992) large-scale land alienation to European settlers between 1881 and 1894, combined with immigration of the Lomwe people from Mozambique in the early 1900s and natural population growth in the southern region, resulted in scarcity of land. Scarcity of land caused major conflicts over land between European estate owners and smallholder encroachers. *Thangata* was developed as a 'solution' to the land problem. Under the arrangement landless Africans were allowed to cultivate small parcels of land on European estates in exchange for labour. The system of forced labour was one of the grievances that led to the 1915 uprising against colonial authorities, which was led by the Reverend John Chilembwe. In 1962, following the country's first multiparty general elections, the Malawi Congress Party (MCP) government abolished *thangata*. The leadership of the MCP argued that the landlord-tenant relationship under *thangata* was exploitative; it effectively turned tenants and their families into slave labour.

In my discussion I have shown that members of wetland communities were accorded various degrees of freedom in what they could or could not do. They also had various obligations to those who allocated the right to use the wetlands. The rights and obligations varied with the mode of access, thereby creating differentiated levels of tenure security among the farmers. Farmers who were not related by kinship to the chiefs and received land from the chiefs could use the land as long as they fulfilled their obligations, namely paying tribute to the chief. Failure to pay the tribute often meant that the chiefs could seize the land. At the time of my research it was reported that four people had lost land in one village because they did not pay their tributes. The numbers could be higher than what was recorded because people cannot speak freely to researchers for fear of reprisals from chiefs (Kambewa 2006).

When these farmers receive land from the chief, it comes with further conditions: they are not allowed to rent it out or pass it on to an heir. They can only pass on the land on condition that the heir is willing to pay the annual tribute. Chiefs have prohibited loaning out and renting out of land because collecting the tribute from renters or borrowers would be more difficult. A final condition requires farmers to cultivate their land constantly, or risk losing it, as happened to an individual who came from Mozambique in the early 1990s, looking for piecework. He was employed as a watchman/herdsman, before obtaining a piece of land from the chief, but it was taken away in 2003 because he did not cultivate during that season since he was sick. The chief took away the land and gave it to those who were able to cultivate and give him bags of rice as tribute. According to the chief, those farmers who did not cultivate their allocated land were lazy, so land had to be taken away and be given to those who needed it the most.

Unlike the situation described above for farmers who have to pay tribute for land received through the chief, the members of the core lineage in the village had fewer obligations. They were only obliged to keep it under cultivation every year. The members of the core matrilineages had secure access and considered themselves to be 'owners of the land'. They freely passed the land to their heirs and loaned it out to other members of the family. Lineage members mostly loaned out the land when they were sick or just to help other members to grow crops. The borrower paid nothing. Some farmers loaned out land in the dry season to others as a way of keeping it clear of weeds for the rice planting season. What is clear is that borrowing was based on trust and it was mostly done among relatives and close friends. The key condition for a borrower was that he or she was only allowed to cultivate the land during the specified period, after which the plot would be given back to the owner. Loans of more than three years were not allowed because it might be tempting to assume that the owner was no longer interested. Unlike those who got the land from the chiefs, a cultivator who borrowed land from ordinary families did not have to pay a tribute.

Regardless of the mode of access no one had the right to sell the land in the wetlands and it was prohibited by the chiefs and their families. In cases where land was sold other family members disputed the sale and claimed the land back. Other disputed 'sales' have taken place, as land has became less available relative to the number of people needing it and those with the ability to raise cash in a desperately cash-short environment were able to make a bid for land for tobacco nurseries in the wetlands. And when considered from a simple notion of supply and demand, it appears these 'sales' of land genuinely resembled market transactions, even though some buyers expressed concern about whether the sales would be valid. At the time no one raised objections on the basis of the land being under customary tenure or questioned the legality of sales. What appeared to be important was that customary procedures had to be followed: the recognised members of the matrilineage had to agree with the impending transaction and the chiefs had to witness the transaction. The chiefs received part of the money from the seller for serving as a witness to the transaction. Recent evidence suggests that this kind of transaction is increasing and is treated as sale. However, in some cases, these transactions have been later redefined as loans, creating a sense of insecurity in the buyer.

5.4 CONCLUSION

The dynamics revealed in this chapter are important and crucial for proper use and management of the wetland resources of Lake Chilwa. If policies and programmes are not sensitive to these dynamics they could create more problems than solutions. Taking these dynamics into account could help promote integrated management. For example, the land policy in Malawi proposes to recognise the common property interest over the wetland for any group of families or individuals living in a locality or having customary land rights in a defined area. The same way of thinking applies to the water policy, which also proposes the formation of catchment management committees to ensure sustainable utilisation of water resource. Furthermore, the irrigation policy also adopts similar thinking and proposes farmer associations and farmer clubs as a pre-condition for supporting irrigation schemes and *dimba* farming respectively (Government of Malawi 2000). The idea is to organise people to manage land and water resources as groups that share common interest and aspirations.

How can government attain commitment to proper management of these resources from a heterogeneous community? Collaboration among various users would be difficult to achieve if access and ownership of the wetland is secure for some people and insecure for others. Failure to understand this heterogeneity risks concentrating the land in the hands of those in power such as the chiefs. If government creates more structures to manage the wetlands, this might intensify competition and conflict over the wetlands with families and chiefs currently having control over them. This calls for policies and laws to be sensitive to the social and power relations existing in the wetland, otherwise there is a chance that the policies will only work to the advantage of the chiefs to accumulate land for themselves at the expense of other villagers.

One would expect little commitment from people who received land from chiefs to invest in proper land use – their tenure status is insecure as they feel they do not own the land. Sustainable management would only be achieved if the chiefs developed rules and used their powers to enforce such rules for proper land management. This suggests that rules and security of tenure, and not ownership, would be the core ingredients of wetland management. In places where members inherit the wetlands from their families there would likely be a higher level of commitment to proper use of the land resources – tenure status is secure because members own the land. In such places it would be important to enhance the concept of ownership.

In sum, this chapter has revealed a localised debate over the value, modes of access and legitimate rights over wetlands, which, so far, has remained invisible to policy makers, planners, and development practitioners. I hope that this detailed analysis of the social and power relations in the wetlands and their relationship to sustainable land management may provide a basis for a framework for the sustainable management of the Lake Chilwa wetlands.

REFERENCES

Government of Malawi. 2001. *Malawi National Land Policy.* Ministry of Lands and Housing, Lilongwe, Malawi.

Government of Malawi. 2000. *National Irrigation Policy and Development Strategy.* Ministry of Agriculture and Irrigation, Lilongwe, Malawi.

Government of Malawi. 2002. *Malawi National Land Policy.* Ministry of Lands and Housing, Lilongwe, Malawi.

Kambewa D. 2006. 'Access and control in Lake Chilwa wetlands: An analysis of social and power relations.' *PhD dissertation* (unpublished). Bunda College of Agriculture, University of Malawi, Lilongwe, Malawi.

Mkandawire RM. 1992. 'The land question and agrarian change in Malawi' pp. 174-175, in: Mhone G (ed) *The Post-Colonial Political Economy.* SAPES Books, Zimbabwe.

Mtika MM. 2001. 'The AIDS epidemic in Malawi and its threat to household food security', *Human Organisation* 60(2): 178-188.

Peters PE and Kambewa D. 2007. 'Whose security? Deepening social conflict over "customary" land in the shadow of land tenure reform in Malawi', *Journal of Modern African Studies* 45: 447-472.

Phiri KM. 1984. 'Yao intrusion into southern Malawi, Nyanja resistance and colonial conquest, 1830-1900', *Transafrica Journal of History* 13: 157-176.

Vaughan M. 1978. 'Kinship and class: Stratification in the Zomba-Chilwa area of southern Malawi, 1800-1914.' *History seminar paper.* Chancellor College, Zomba, Malawi.

Chapter 6

WATER USE AND CONFLICTS IN SOME RIVERS IN LAKE CHILWA BASIN

Wapulumuka O. MULWAFU

6.1 INTRODUCTION

Malawi is usually described as a country with abundant water resources. In a large measure, this is true on account of the existence of many rivers and lakes that crisis-cross the country. Compared to many arid and semi-arid countries in Africa, Malawi is endowed with water resources which, if properly harnessed, could reduce water shortages and improve the lives of a large segment of the population. However, a closer scrutiny shows significant challenges in accessing and using the water resources. Although 3 000 m^3 of water per capita are renewed annually in rivers and lakes, only 300 m^3 per capita are available in the dry season. This means that a fairly small proportion of water resources in the country are available for the various domestic and productive uses. Moreover, significant variations exist in the distribution of, and access to, these water resources. This scenario raises a number of important issues about water resources use and management in Malawi.

One of these issues concerns the shift in the perception of water as a social good that used to be freely available to every person to that of a precious, scarce resource whose control and ownership is increasingly being contested. When its supply is limited in quantity or quality or its distribution is uneven, water can be both a source of co-operation and contestation among its different users. Since different groups of people have different sets of needs and values, access to scarce water resources becomes a competitive process requiring people to exercise their various coping strategies and survival skills.

In this chapter I argue that although water supply in Lake Chilwa is generally sufficient, its distribution and use is extremely skewed. This creates conditions for conflicts that underscore the deeply entrenched power relations existing in the area, particularly between the primary and productive users of water resources. Primary users of water are usually small-scale farmers who use water to meet such as drinking, cooking, and washing while productive users are people who use water for commercial purposes such as irrigation, tobacco production and brick making. Power is wielded by a range of people such as the original inhabitants of the area, the owners of private water sources, and community leaders. Conflicts have also increased as a result of the emergence of many development project interventions in the area. Lake Chilwa is by far the best documented basin in the country, with studies dating as far back as the 1960s. Most of the studies have been on the biological diversity of the catchment area and the waters of the lake itself (Kalk *et al.* 1979, Dolamore 1987, Petermann 1990). Thus the availability of longitudinal data sets allows us to analyse developments that have taken place over time. By highlighting the socio-cultural dimension of contestations over water use, this chapter is intended to contribute to the growing body of literature on water resource management in the country. Although I have tried to cover issues relating to the entire Lake Chilwa basin, the major focus is on water resource management around the lake itself and the two rivers of Likangala and Domasi.

6.2 METHODOLOGY

The chapter is the product of a wider research study on the socio-economic aspects of water resource management conducted in the Lake Chilwa catchment in southern Malawi between 2000 and 2006. The basic method of research was an extensive and critical review of policy documents and relevant literature on water resources in Malawi. I conducted the review to discern what is already known about the subject matter and so avoid repetitions, to establish gaps in knowledge that may need to be addressed and to unravel other pertinent research issues. The second method involved baseline surveys, which I carried out to obtain additional information from water users in the area. To collect data from key stakeholders, I designed and administered a set of questionnaires on water sources, irrigated agriculture, income, and food security to farmers and water users, who were purposefully sampled. In addition to surveys, field assistants who were resident in the research sites collected qualitative information. They used basic social science methods such as focus group discussions, key informant interviews and field observations, and sought information on the causes and nature of conflicts over water, as well as participants' perceptions of water pollution.

6.3 USE AND MANAGEMENT OF WATER RESOURCES

Apart from the lake and the rivers, I identified a range of water sources in the study area, which included streams, rainwater, protected and unprotected wells, small dams, boreholes and taps. However, these water sources are not evenly distributed across space, with less water in the largely rural communities located in the downstream areas and more water in the upstream areas of Mulanje Mountain and the Zomba Plateau and in areas adjacent to urban concentrations. In addition, the upstream areas have more boreholes and tapped water sources than the downstream areas. Downstream communities rely mostly on the lake itself, the river, wells and a few boreholes.

The 1986 Water Resources Master Plan described water allocation in terms of three major activities: domestic water supply for urban development, irrigation schemes and fisheries production (Government of Malawi 1986). In this study area, water is used for a variety of purposes but the main ones are:

– drinking by people and animals;
– domestic purposes, such as cooking, bathing and washing;
– small-scale *dimba* (stream bank) farming in the wetlands;
– large-scale rice irrigation schemes; and
– large-scale estate production.

Each water source has rules and rights that govern its use by the communities surrounding the source. While some of these rules appeal to the natural rights of human beings to use water resources, others are clearly made by the people themselves to deal with existing conditions in the area. For example, water in the Likangala and Domasi Rivers is generally seen as a gift from God to His creation.[1] Thus, every person has a right to use the water for any purpose. For instance, if water passes through an individual's land, that individual may divert the course of the water to construct a dam or use it for irrigation. However, there is a tacit understanding that in using water from the river, all users need to be responsible enough to consider the needs of other people (Simpson 1999). Although water from these perennial rivers is generally adequate for everyone's needs, it is not always clean or safe for drinking purposes.

Wells are governed by some expressed rules. Often owned and controlled by the individuals who dug them, communities can access the water for domestic purposes only. Except for the owner, no one can use the water for productive purposes such as irrigation or washing clothes or dishes in the vicinity of the source. For example, a farmer/businessman at Ramusi Trading Centre who owns a covered well does not allow his neighbours to use the source for activities other than domestic ones. He does not ask other users to pay for water and yet he is responsible for the maintenance of the well.[2] Wells can be unsafe, especially if no sanitary measures have been made to protect them. In cases where there is no proper protection and, where, for instance, foreign objects have fallen inside, it can be difficult to remove them; yet the entire burden falls on the shoulders of the owner, who may occasionally receive help from friends.

Another important source of water is the Muliyankhwazi reservoir in Chiliko Village, which provides water to the general public throughout the year. Tradition holds that this small dam was created naturally and has been in existence for hundreds of years. The reservoir contains fish that attract people from the surrounding villages to come for fishing. Most people use water from this source for bathing and washing. Everybody is free to use water from this source. But in the dry season, the local residents tend to experience competition from commercial farmers who divert water for their irrigation schemes at Likangala and Chiliko.

1 Interview: Mr S. Nkhota (Village Headman Chiliko) Irrigation Sector dated 2 March 2000.

2 Interview: Mr Thunya, Ramusi Trading Centre, Irrigation Sector dated 16 February 2000.

Boreholes and taps are often provided by government through self-help schemes or by non-governmental organisations such as churches. Ideally, these sources are supposed to provide safe and clean drinking water. They also have the capacity to provide enough water when the sources are working perfectly although, based on experience in the last few years, less than half of them function properly for more than six months in a year. In addition, more than 30% of boreholes are dysfunctional at any one time of the year. The government's recent policy has been to empower local communities in the operation and management of such water sources. To this effect, committees have been established that set up their own rules for the use and management of the water source. Some of the commonest rules governing the use of boreholes and taps include:

- Users (members) must make a regular contribution of money to maintain the water source.
- Drawing water is done on a first-come first-served basis.
- Cleaning of the water source should be done regularly.
- There should be no washing of utensils and nappies near the source.
- Each household can only draw one pail of water at a time.
- Children should not draw water on their own without the presence of an adult.
- Non-members cannot draw water unless prior permission has been sought.

In principle, all water points are supposed to have local committees to administer their use and management. The roles of these committees are, among other things, to raise money for the maintenance of a water source, to ensure general cleanliness, and to discuss any other issues affecting the water sources.

Domestic water sources are provided by the individual users, the government and non-governmental organisations, with different requirements and placing different responsibilities on the users. People have a variety of use rules for these sources. Skjolsvold (2010) has examined the challenges of introducing new institutional arrangements in the management of water sources that do not take into account pre-existing social structures. Using the case of two water sources in the Kasungu area of Malawi, the author shows that such interventions do not necessarily result in their success or sustainability. In another earlier study of boreholes in Malawi, de Gabrielle (2002) observed that the provision of more water sources may not necessarily lead to an improvement in peoples' lives. For example, in Malawi more than 30 000 boreholes have been drilled since the 1980s and yet more than 30% were dysfunctional by 2004.

The supply of water also has a political dimension. Politically and economically powerful groups tend to control access to water sources for their own interests (Cleaver 1995, Peters 1984). In colonial Botswana and Zimbabwe, for example, water supply became a political weapon for the elite communities. In Malawi, water supply has not been spared from politics either. In the run up to the 1999 general elections, the government decided to construct more than 5 000 boreholes across the country, most of which did not even meet the required minimum standards. This was an apparent attempt to woo rural voters and since then it has become almost routine for Members of Parliament to request the line ministry to construct boreholes in their constituencies. It seems the provision of water supplies has become an important element in the strategy to win electoral votes and maintain political support.

6.4 THE USE OF WATER FOR DOMESTIC AND PRODUCTIVE PURPOSES

The provision of water to meet the minimum requirement for domestic purposes is obviously necessary to reduce the incidence of diseases and to improve the lives of people. Although Malawi has a per capita domestic water use of 29.7 litres per person per day, my study suggests that the figure may be even lower for the Lake Chilwa area. At Kachulu harbour, residents pay a high price for water drawn from wells and boreholes, ranging from K50 to K100 (US$0.30-0.75) per 20 litre bucket.[3] This is considered high relative to the fact that many people do not generate their own incomes. But they are forced to pay in part because of problems of water scarcity. The low land areas of the wetland receive only between 800 and 900 mm per year (Government of Malawi 2000), rainfall is very erratic and unreliable, and water sources are few and far apart, as can be seen in Table 1 below.

Another issue emerging from my study is the prioritisation of the use of domestic water supplies. Most of my respondents mentioned that, in order of priority, the following are the key uses: domestic human needs, including drinking, cooking, bathing and sanitation; beer making; livestock watering; gardening; and brick making.

Table 1: Water points in the Likangala catchment area

Research sites	Covered wells	Open wells	Boreholes	Taps
Urban	2	2	5	2
Peri-urban	0	1	4	3
Estates	3	5	2	2
Irrigation	4	8	3	0
Fisheries	3	3	2	0
Total	**12**	**19**	**16**	**7**

Source: BASIS (2001)

3 BASIS Project Research Notes, dated 28 May, 2001. Data collected by Noel Mbuluma.

Table 1 above shows that the upstream area (urban and peri-urban sectors), depends on taps and boreholes. On the other hand, the downstream area (irrigation and fisheries) uses boreholes and wells. Water shortage is a common occurrence in the downstream area. The table also gives an indication of the state of water sources in the Likangala River catchment. For an area of about 100 villages and a population of over 80 000, this is a very small number of water sources. This does not include the population in the Municipality of Zomba, most of whom are supplied with piped water facilities. But for most people living along the lower part of the Likangala River, which is also mostly rural, the river is their major source of water.

Scholars in other contexts have dealt with the issue of water use. For example, Cleaver has bemoaned the persistent use of the dichotomous classification of water into productive and domestic sub-sectors, arguing that more often than not, this division does not help much in understanding the power relations involved in water resource management (Cleaver 1995 and 1998). In addition, the approach is so technocratic that it tends to focus on formal user groups, and in the process leaves out women's participation. Traditionally, women have been confined to domestic use of water while men have tended to dominate the formal water user groups. Cleaver argues that a gendered approach would underscore the role of women in water resource management and their efforts in alleviating their poverty. By focusing on water as an economic good, researchers and policy makers tend to underestimate the importance of domestic uses. This is because the economic benefits of productive water can be easily quantified, while the health and social impact of improved domestic supplies are difficult to measure.

Water is used for a variety of productive activities throughout the Lake Chilwa basin and these include brick making, dry-season gardening, livestock watering and beer brewing. Studies of domestic uses of water supplies must incorporate issues of power relations, as they affect access to such water resources. As this study shows, although most water sources have committees with rules and regulations restricting the use of domestic water for productive purposes, political and socio-economic factors influence how and why some individuals can use a scarce resource for productive purposes.

6.4.1 Brick making

Most brick making is done along the major rivers and streams in the area although this activity also takes place in other areas as well. Sometimes people use water for brick production from taps, wells, boreholes and irrigation canals. The bricks are usually not used locally for building houses. They are sold to urban and peri-urban areas where a thriving building industry exists and better prices are offered. In June 2001 the price of bricks was K600.00 per 1 000 burnt bricks and this included labour charges for loading the bricks into lorries. In 2010, the same number of bricks were selling at K3 000. The competition for buyers is intense. This situation sometimes leads to people selling bricks at give away prices as buyers take advantage of the poorly organised brick makers.

Brick making is an important alternative source of income for local people. Moreover, it is not an exclusive activity of men; women also take part in the enterprise. However, it is interesting to note that most of the women involved in this activity are family heads or widows. Informants observed that family responsibilities and poverty are the key factors for their participation in brick making activities.

Brick making has both a positive and negative impact on the environmental status of the area. It contributes greatly to soil erosion and pollution of the rivers in the Lake Chilwa catchment. It has also resulted in deforestation, erosion gullies and the removal of top fertile soils in the Lake Chilwa basin. An interesting aspect is that people prefer using soil from anthills for brick making because it is less sandy. But anthills are also used for planting crops like maize because they are considered to be fertile areas. Such competing demands underscore the fact that resources are not always used for a single purpose but rather that individuals and communities make decisions depending on their own interests.

6.4.2 Vegetable gardening

Dry season gardening is another important economic activity that uses water resources. In both urban and rural areas, people plant maize and a variety of vegetables for home consumption and for sale. In urban areas it is common to see gardens around houses of residents. Away from urban concentrations, people often use water from wells and boreholes to cultivate vegetables, even though water point committees do not allow this. These people usually own gardens close to the water sources or have some disproportionate amount of influence in the area, as is the case with most *dimba* owners along the Likangala River, and it is a potential source of conflict.

Dry season cultivation is a major livelihood activity in the study area. It shapes social relations and helps in alleviating poverty. Access to *dimba* gardens is determined not just by membership to a local community but other factors such as marriage relationships and the ability to make indirect forms of payment (Peters and Kambewa 2007). There are hundreds of *dimba* gardens in the study area with differing sizes and shapes, ranging from 0.01 hectare to 0.5 hectare per household. *Dimba* cultivation is a widely adopted, productive use of water in the informal sector and one that offers lasting opportunities for the improvement of living standards of peasant farmers in the country. However, despite the significant role that *dimba* cultivation plays in the rural economies, it is not recognised in the government's new water and irrigation policies (Government of Malawi 2005, 2000). Moreover, it conflicts with the current environmental laws and policies, which prohibit the cultivation of stream banks. The Department of Environmental Affairs prohibits cultivation of areas less than 8 m from river courses as part of the effort to arrest environmental degradation in fragile areas (Government of Malawi 1996).

6.4.3 Livestock watering

In general, Malawi does not have a big livestock industry. Compared to other countries in the southern African region such as Botswana and Zimbabwe, the population of livestock is relatively small and this means that competition for water between livestock and human use is less intense. But given the scarcity of water in the Lake Chilwa area, access to water for livestock often competes with other activities and this is a potential source of conflict. The most commonly kept animals are goats, cattle and chickens. In urban areas, some people keep broilers and laying chickens for commercial purposes. These fowl consume large quantities of water each day and they mostly require treated water. In rural areas, livestock range free. Goats especially are left to stroll to water sources and as long as they do not damage the sources most people seem to have an indifferent attitude. In terms of policy, Malawi recognises the provision of water for livestock as a domestic activity.

6.4.4 Beer brewing

Beer brewing is another major economic activity that uses water. In particular, *kachasu* (a local gin) is commonly produced in the area and it is mostly a woman's activity. Although brewing is often done during the dry season and particularly after the harvest of the wet season crops, beer brewing and drinking is an important social function. It is also an important means of earning income for those people who undertake it. Additionally, the production of *thobwa* (a fermented sweet drink) is common during this period as well. Women vendors are often seen selling their products to the public at the intersections of strategic pathways and market centres.

6.5 CONFLICTS OVER WATER USE

The regulation of river flow regimes and the conflicts it generates is a subject that has been extensively studied elsewhere (Adams 1990, 1992). These studies have shown that most of the conflicts tend to revolve between users in the upstream and downstream areas. In my case, the study of water use and management by communities living along the Likangala River is similarly replete with stories of conflicts and tensions that operate at different levels. There are manifest as well as latent conflicts. Underlying these conflicts is the fact that water has increasingly become a scarce resource. The seasonal variations in the supply of water and the increasing demands for it in the upstream area create many problems for the users.

In the past, people were able to access water within a short walking distance. Now they not only have to walk long distances to fetch water but they also have to wait in queues for a long time. The real problem is obviously the waiting time at the various water sources. Many respondents complained that people could wait for up to three hours in a day to get one *chigubu* (20 litre container) of water.[4]

The use of water resources in the study area is gendered in ways that suggest an inequitable distribution of scarce resources: more men use water for productive purposes while women's use of the resource tends to be restricted to primary activities. This means that women, and particularly female-headed households, are less empowered to access water resources for productive purposes. Moreover, the scarcity of water which requires women to spend long hours waiting to fill their containers means that they have less time to invest in other productive activities. More empirical research needs to be done to establish the impact of this demand on time on the income levels for women, particularly female-headed households. Studies elsewhere have shown that new policies and practices, which place additional demands on women, have continued to displace them from the management of water resources (Cleaver and Elson 1995). Participation in water management is also gendered in the sense that more women dominate membership of water point committees while men are actively involved in decision making at the higher management levels.

Some people contest the notion of 'primary' or 'domestic' use of water. Users also question what constitutes a 'domestic' purpose. For example, while drinking, washing and cooking are commonly accepted as domestic, the use of drinking water by livestock is not. Sometimes washing utensils and nappies is also not considered part of domestic use. There have been conflicts over who should use the water for what purposes at various water points.

In other cases, users contest the notion of the 'right' to use water. Traditional values hold that water is a gift from God. While it is accepted that every person has a right to use water, some variations do exist, depending on particular circumstances of the water source. It is also evident from my study that some people have more rights than others. A recent immigrant into the area may have fewer opportunities to use a particular water source than an older settler. Rights to water use are also dependent on wealth, gender, status and age. A research assistant stationed in the Estate Sector reported that "on my countless visits to draw water at the taps, I have always been given the first seat ... At first I was startled. I therefore sought the views of Mrs Ellen Kainga. She said, 'It is rare to find a man drawing water ... so he is given the front seat'".[5] In one village, residents refused to allow their neighbours to use a borehole on account of the fact that the residents have to pay for its maintenance. They further argued that if their borehole would break

4 BASIS Project Research Notes, dated 30 May 2002. Data collected by Davidson Chimwaza.

5 Basis Project Research Notes from the Estate Sector dated 11 March, 2000. Data collected by Davidson Chimwaza.

down after being overused, their neighbours would not help them with money for its maintenance.

Conflicts sometimes occur at the intra-household level. Impatience, selfishness and petty jealousies within or between families sometimes extend to bullying or ostracising others when accessing water sources. In some cases, users simply ignore the rules and capitalise on intra-household differences to get water before the others on the queue. One informant, who owns an open well at Ramusi 2 in the Irrigation Sector, noted that 'there have been times when women fight for water here, specifically during the dry season when the water level goes down and one has to wait [for a long time] to fetch water'.[6] Petty jealousies and suspicions occur frequently, as is exemplified by an incident that occurred in December 1999 at a water tap at the Mikuyu Prison Museum. A certain woman decided to get to the head of the queue. But when obstructed, she began cursing all the women at Mikuyu Prison Museum houses. She is quoted as saying, 'I know there are some women at the museum who don't want us to draw water from their taps. But one day we will see. I will beat one of them and that will serve as a lesson to the rest.'[7] This incident soured relations among women residents in the area. What started as a simple matter became serious with counter accusations involving issues of arrogance and witchcraft.

There have been reports of school children contesting with women (domestic users) over access to drinking water. This often occurs in areas where a water source, such as a borehole, tap or well, is located near a school. During break time students rush to drink water from the source. Because of time limitations, they tend to jump the queue so they can drink and quickly return to classes. However, this breaks the first-come first-served principle set down by the water point committee.

The positioning of water sources is another bone of contention. Where and how boreholes and taps are sited often depends upon several factors, such as water level and proximity to houses. Given the scarcity of water in the area, everyone would like to have the water source close to their homes. But donors are ideally supposed to place the sources in areas of high water table. However, this is not always the case. A clear case comes from the peri-urban sector where conflicts have arisen over the siting of boreholes. At Mteche Village a borehole was sunk at a place of low water table, hence it is often non-functional. The siting of a borehole at Mainuka Village is believed to have been politically motivated. The former Member of Parliament for the area had apparently had it located there to win parliamentary votes. When he lost in the 1999 elections his supporters vandalised it and stole the equipment to express their displeasure at the apparent show of 'ingratitude'.

6 Interview: Mai Maggie Matope, Irrigation Sector dated 17 May 2000.

7 Interview: Mai Christina Magalasi, Estate Sector (Notebooks) dated 25 December 1999.

6.5 CONFLICTS OVER WATER USE

The establishment of agricultural estates and irrigation schemes along the Likangala River is another source of contestation for water use. Some of the major commercial enterprises in this region include the Likangala irrigation scheme complex, Domasi and Khanda irrigation schemes and the Chimpeni and Kachulu estates. While irrigation schemes mainly produce rice, estates produce tobacco for commercial purposes. Under the irrigation schemes, more men tend to access plots than women and this has implications for understanding the nature of conflicts over water use. Since these agricultural activities require a huge and regular supply of water, dams and canals have been dug to divert water from the natural course of the river. This creates problems for people downstream. During the wet season, the canals serve as conduits for the water to flood downstream villages. However, when the water level goes down in the dry season, large scale farmers, who desperately need the water for the irrigation of their crops, divert its natural course. An example is protracted conflict between water users at Kalinda village in the irrigation sector. Communities living in the downstream area complained that they were deprived of adequate water during the dry season because the water had been diverted upstream for irrigation purposes at the rice scheme. The introduction of the scheme increased the incidence of flooding downstream because a canal was dug to redirect the flooding downstream towards this particular village.[8]

The flooding or lowering of the water level in the river also engenders conflicts among the different users. These seasonal changes have in some cases been induced by human activity: informants observed that the reduction in the flow of the river has been caused by deforestation of watershed areas. They noted that deforestation has increased greatly in recent years because people use trees for building houses, firing bricks, making canoes and making fires for drying fish. The result is that water flow has been declining, thereby making it difficult for communities downstream to have sufficient supplies of water for productive activities.

Lake Chilwa is another important source of water for the communities living downstream. Although the water is not used for drinking purposes, it is needed in other productive uses and thus generates conflicts among the users. Fishers blame upstream users for the reckless cutting down of trees and poor agricultural practices that have resulted in soil erosion and the eventual siltation of the lake. The siltation in turn results in low fish catch by the fishing communities in the area. Men comprise the majority in these fishing communities and control most of the processes while women only come in at the point of transporting and selling fish.

Dambo areas are also important sites of conflicts. The importance of *dambos* comes from two main factors. First, due to population pressure, there is a serious

8 Interview: Mr Augustine Fatch, KalindaVillage, Irrigation Sector dated 7 December 1998.

land shortage problem in the area. The population density has increased from 162 people per km^2 in the 1980s to 321 in 2008 (Government of Malawi 2008) which is quite high compared to the national average of 104. Second, even where people have land, the rainfall is often inadequate and unreliable so that cultivation in the upland areas does not produce high yields. *Dambos*, therefore, remain by far the best and surest way of producing such food crops as maize, peas and potatoes. But *dambo* land is relatively inadequate to support the population and their various productive activities in the area. Consequently, some people have resorted to cultivating their crops on stream banks. Cultivation of stream banks is a clear violation of government policy on the environment. Mr S. Mwandama of Ramusi 2 in the irrigation sector has reported conflicts between farmers and the environmental staff over the cultivation of land along stream banks. The agricultural and environmental staff recommend that there should be no cultivation at least 8 m from the river but most farmers usually ignore this rule.[9]

As stated earlier on, livestock production is not a major economic activity in the area so few people keep domesticated animals. The commonest animals are goats and chickens although a few households also keep pigs, sheep, ducks, turkeys, doves, guinea fowls and rabbits. Goat and cattle owners experience conflicts over access to grazing land such as the dambo areas in the dry season. Wetlands are not only ideally suited for a range of agricultural crops but they also have good grasses for livestock grazing. But allowing livestock to graze in the wetlands would result in the destruction of crops in the fields. There is also another problem regarding whether livestock should drink water in the river upstream when some people use the same water for domestic activities downstream. People complained that livestock contribute to the pollution of the water. They also pointed out that some people have the tendency of allowing their livestock to drink water around a borehole or a tap, which makes the water source dirty.

There is sand mining at different points along the rivers in the Lake Chilwa basin. Not only does this activity pollute the water but it also contributes to the destruction of river banks, thereby altering the natural flow of the rivers. Downstream residents experience problems with the muddy water. In addition, settlements in the upstream area, such as Chikanda, St. Mary's and Mpondabwino contribute to the pollution of the Likangala River. A good amount of refuse from households in these locations is often dumped in the river.

The increased demand for water resources and the growing inequalities in access to such resources have intensified the level of conflicts in this area. Consequently, people have developed various coping strategies as well as means of resolving these conflicts. At the individual and household levels, people discuss problems that are not very serious, such as quarrels over who should draw water first from the source. Heads of households are asked to help resolve the con-

9 Interview: Mr S. Mwandama, Ramusi 2 in the Irrigation Sector dated 4 March 2000.

flict. However, if a problem is deemed too difficult to be handled by the individuals themselves, the intervention of a local water source committee is sought, where this exists. For example, water committees deal with matters relating to non-payment of money for maintenance of a water source or failure to clean it up. All serious matters are reported to the chiefs who, depending on the gravity of the problem, may involve a village headman or a traditional authority. Chiefs are supposed to handle all cases that cannot be resolved by the water committees in the different villages such as disputes among families. But criminal cases are reported to the police, such as vandalism of a water source or fighting that results in injuries.

The increasing use of water for industrial and other productive activities in the upstream area creates numerous problems for downstream communities. It also creates conditions for potential conflicts. For example, the establishment of an abattoir near the Likangala Bridge, and the leakages from the sewage disposal systems of Zomba Central Prison, Zomba Central Hospital, Cobbe Barracks and Zomba Mental Hospital has resulted in the dumping of waste pollution into the river. Mai Roseby Nandolo remembers that a few years ago 'a vehicle moved around the village announcing that we should stop drinking and bathing from the Likangala River because Zomba Central Hospital had emptied sewage into the river. Since we had nowhere to draw water, we continued taking water from the river'.[10]

The decline in water quality is often attributed to pollution by various agents and most people are aware of the fact that the water they use is polluted. A focus group discussion with water users at Ramusi 1 in the irrigation sector revealed that 'Likangala River is the worst source ever since. Every year people suffer from cholera just because of this source'.[11] Another informant said that it was not uncommon to see children using the river as a toilet.[12] Downstream users lamented that the colour of the water has changed to green and there is a lot of debris and other rotten material seen floating in the river. The water quality becomes particularly bad during the wet season when a lot of foreign objects are washed down into the river and at that point its colour changes completely. Some people have seen dead snakes and frogs, which further contribute to the pollution of the water sources.[13] Another problem is that every user/household has their own bucket for drawing water and it is difficult to determine the hygienic condition of these buckets. Finally, an open well at Nansongole in Chiliko Village is located near the graveyard and there is no guarantee that the water quality is good at all. Consid-

---

[10] Interview: Mai Roseby Nandolo, Estate Sector, (Notebook) 25 January 2000.

[11] Focus Group Discussion with Mrs E.Unyolo, Mrs M. Kang'oma, Mrs F. Walley, Mrs D. Nyalape, Mrs E. Katuli, Mrs E. Tandwe of Ramusi 1 in the Irrigation Sector dated 12 May 2000.

[12] Interview: Mr G.C.K Chiwaya, Irrigation Sector dated 12 February 2000.

[13] Interview: Mai Nanthambwe, ChilikoVillage, Irrigation Sector dated 10 February 2000.

ering that there may be a possibility of the water being polluted, an official from the Health Department advised the people to construct another well far from the graveyard. That has not been done and, meanwhile, people continue to get water from the same well.

The pollution of the river has obviously created health hazards for the riparian communities, with an increase in cases of cholera, bilharzia, dysentery, typhoid fever and other water-borne diseases. These cases are confirmed by a water quality study that showed a high presence of faecal coliform and faecal streptococci in the various water sources (Chavula and Mulwafu 2007). Skin diseases too, such as scabies, also occur on a regular basis. Mai Nandolo further complained that at one time we heard that prisoners in Zomba had developed scabies yet in that condition were being asked to bathe in Likangala every Mondays and Fridays. It was rumoured that those villagers who would continue bathing from the river would contract scabies.[14]

Although it is common knowledge that the Likangala River is heavily polluted and, although the riparian residents would prefer to avoid using it, they are left with absolutely no choice. They have to use the polluted water and live with the consequences. As one informant complained, 'The most popular source is the open well at the Agricultural Office, though in true sense it is the Likangala River. It is the most popular point just because it supplies water all the year round for about six villages. I am sure that very soon you will be able to see [. . .] long queues.'[15] Given the gravity of water problems in the area, another informant despaired that 'there is nothing to be done since this is to do with nature. But what I can say is that people should pray hard to the Lord'.[16]

There are conflicting views on the general public's perception of the time of year when the water quality may be considered good. Some people said that water quality is good during the dry season because, although it tastes salty, the colour is generally clear and it does not harbour many water-borne diseases. On the other hand, other people said that the water quality is good during the wet season when the water runs fast and filters the dirt. The only problem is that it is muddy and tends to carry many dirty objects.

Transboundary waters are another potential source of conflict (Ohlsson 1995). Since the Lake Chilwa basin shares a boundary with Mozambique, it is important that a transboundary approach be adopted in the management of water resources. But, as I have noted earlier on, little information is available on the Mnembo River and other water sources on the Mozambican part of the catchment. In order

---

14 Interview: Mai Roseby Nandolo, Estate Sector, (Notebook) 25 January 2000.

15 Focus Group Discussion with Mrs E. Unyolo, Mrs M. Kang'oma, Mrs F. Walley, Mrs D. Nyalape, Mrs E. Katuli, Mrs E. Tandwe of Ramusi 1 in the Irrigation Sector dated 12 May 2000.

16 Interview: Mr S. Nkhota (Village Headman Chiliko) Irrigation Sector dated 2 March 2000.

to achieve efficiency and sustainability, an integrated approach would require that all stakeholders be involved in the use and management of water resources.

6.6 CONCLUSION

Conflicts in the use and management of water in the Lake Chilwa basin is but one example of the larger problem experienced by rural communities throughout the country. This study shows that although water resources are sufficiently available, their use and management has become an important site of negotiation and contestation by the riverine communities. The traditional idea of right of use of water resources is rapidly being contested as the number of productive users increases.

What counts as domestic use is relative. Even where minimum water requirements are not met, users venture into productive activities because that is seen as a possible way of generating income and getting out of the poverty trap. If poor people's access to water is limited their opportunities for poverty alleviation also decline. The traditional categories of domestic and productive use tend to be frozen in time and space in the sense that they do not allow for negotiations that take place among poor people.

The provision of water facilities to rural people must be seen not only as a means of improving their health and sanitation conditions but also as a way of ultimately alleviating their poverty. Water plays a vital role in the social and economic activities of rural people. Driven by the need to improve their lives, people use domestic water for productive purposes in ways that may not be easily comprehensible to observers.

REFERENCES

Adams WM. 1990. *Green Development: Environment and Sustainability in the Third World.* Routledge, London.

Adams WM. 1992. *Wasting the Rain: Rivers, People and Planning in Africa.* University of Minnesota Press, Minneapolis.

Chavula G and Mulwafu WO. 2007. 'Hazardous water: An assessment of water quality and accessibility in the Likangala catchment area in Malawi', *Malawi Journal of Science and Technology* 8: 30-41.

Cleaver F and Diane E. 1995. 'Women and water resources: Continued marginalisation and new policies.' *Gatekeeper Series No.49.* International Institute for Environment and Development, London.

Cleaver F. 1998. 'Choice, complexity, and change: Gendered livelihoods and the management of water', *Agriculture and Human Values* 15(4): 1-115.

Cleaver F. 1995. 'Water as a weapon: The history of water supply development in Nkayi District, Zimbabwe', *Environment and History* 1: 313-33.

DeGabriele J. 2002. *Improving Community-based Management of Boreholes: A Case Study from Malawi.* WI: BASIS Progress, Madison.

Dolamore JA. 1987. *The Water Balance of Lake Chilwa.* Department of Geography, University of Loughborough.

Ferguson AE and Mulwafu WO. 2007. 'If government failed, how are we to succeed? The importance of history and context in present-day irrigation reform in Malawi', pp. 211-227, in: Van Koppen M, Giordano M and Butterworth J (eds) *Community-based Water Law and Water Resource management Reform in Developing Countries: Comprehensive Assessment of Water Management in Agriculture, Series 5*. CAB International, Wallingford, UK.

Ferguson AE and Mulwafu WO. 2004. *Decentralisation, Participation and Access to Water Resources in Malawi.* Madison, Wisconsin and Development Experience Clearing House, USAID. Available at:http://www.basis.wisc.edu (Accessed on 20 February 2011).

Ferguson AE and Mulwafu WO. 2004. *Irrigation Reform on Malawi's Domasi and Likangala Smallholder Irrigation Schemes: Exploring Land-Water Intersections.* BASIS CRSP, Madison, Wisconsin.

Government of Malawi. 2001. *Lake Chilwa Wetland State of the Environment Report.* Ministry of Natural Resources and Environmental Affairs, Lilongwe.

Government of Malawi. 1986. *National Water Resources Master Plan.* Ministry of Natural Resources and Environmental Affairs, Lilongwe.

Government of Malawi. 1996. *The Environmental Management Act of 1996.* Environmental Affairs Department, Lilongwe.

Government of Malawi. 1998. *State of the Environment Report for Malawi of 1998.* Environmental Affairs Department, Lilongwe.

Government of Malawi. 2000. *National Irrigation Policy and Development Strategy.* Ministry of Agriculture and Irrigation, Lilongwe.

Government of Malawi. 2005. *National Water Policy.* Ministry of Irrigation and Water Development, Lilongwe.

Government of Malawi. 2008. *Population and Housing Census.* National Statistical Office, Zomba, Malawi.

Kalk AJ and Howard Williams M. 1979. 'Lake Chilwa: Studies of change in a tropical ecosystem', *Monographiae Biologicae*, 35:17-41.

Ohlsson L. 1995. *Hydropolitics: Conflicts over Water as a Development Constraint.* Zed Publications, London.

Petermann T. 1990. *Report on the Environmental Impact of Self-help Rice Irrigation Schemes around Lake Chilwa (Malawi).* Report submitted to the Liwonde ADD, Government of Malawi, Ministry of Agriculture, Malawi.

Peters P. 1984. 'Struggles over water, struggles over meaning: Cattle, water and the state in Botswana', *Africa Affairs* 54(3): 29-48.

Peters PE and Kambewa D. 2007. 'Whose security? Deepening social conflict over "customary" land in the shadow of land tenure reform in Malawi', *Journal of Modern African Studies* 45(3): 447-472.

Skjølsvold TM. 2010. 'One size fits all? Institutional design: Lessons from two Malawian examples', *International Journal of the Commons* 4(2): 758-771.

Simpson M. 1999. 'Local knowledge management systems and the procurement of domestic water supplies in rural Malawi', pp. 111-130, in: FitzGibbon JE (ed) *Advances in Planning and Management of Watersheds and Wetlands in Eastern and Southern Africa*. Weaver Press, Harare.

Chapter 7

RE-EXAMINING LAKE CHILWA BASIN GOVERNANCE CHALLENGES: AN INSTITUTIONAL PERSPECTIVE

Blessings CHINSINGA

7.1 INTRODUCTION

The current re-examination of the governance challenges facing Lake Chilwa's floodplain has been inspired by the overwhelming consensus that it is under enormous pressure from people from both within and outside the area to support their livelihoods (Peters 2004; Kambewa 2004; Njaya 2009). The floodplain is a destination for people from both Malawi and neighbouring Mozambique. They consider the area as a saviour of last resort because of great variations in climatic patterns which are negatively affecting the livelihoods of the people elsewhere while livelihoods on the plain appear to be resilient. The richness of its natural resources and the ease of gaining access to them in the absence of proper regulation since 1980 make the floodplains an attractive destination (Mvula and Haller 2009). Consequently, the enormity of the pressure has transformed the floodplain into a site of widespread conflicts and contestation between different stakeholders regarding access, use and management of its various valuable resources. This has, in turn, created a situation in which the threat of the floodplain being degraded is always imminent.

While the floodplain has a long history of resource exploitation stretching to the pre-colonial era, it only became a centre of focus for scholarship and policy debates in the early 1970s, which intensified at the start of the 1990s. In the early 1970s scholarly research of the Lake Chilwa floodplain was led by Margaret Kalk who served as a Professor of Biology at Chancellor College, University of Malawi. She conceived, inspired and later synthesised into a book a multidisciplinary study about the plain, which focused on the dynamics of change from various perspectives (Kalk *et al.* 1979). Consequently the challenges associated with the management of the wide range of valuable resources on the plain, particularly fishing and land for irrigation, could no longer be ignored. In biological terms, the carrying capacity of the plain was quickly being stretched to breaking point. Contestation over access, use and management of valuable resources became widespread.

The intensification in demand for the floodplain's resources may be attributed to several factors. The establishment of irrigation schemes in the area has dramatically increased the value of the land associated with increased demand for foodstuffs in the surrounding urban centres. The wetlands and stream bank gardens have become even more valuable in the wake of regular droughts as sources of income and food (Peters 2004, Kambewa 2004). The demand for fishing in the floodplain has also grown dramatically, triggered mainly by the deleterious effects of the economic crisis of the 1980s. As the implementation of government's structural adjustment programmes forced people out of employment, many people opted for fishing as an alternative form of livelihood. The pressure on the plain continued to mount as an increasing number of fishermen migrated from other water bodies that had been overexploited, hoping to revive their fishing occupations (Donda 1997; Njaya 2009).

The declaration of the plain as a Ramsar Site in 1997 has further elevated its significance and visibility in scholarship and policy circles. In the last two decades, the floodplain has been a subject of various studies and policy initiatives that have sought to provide a better understanding of the social, economic, political and even cultural dynamics about access, use and management of the various resources therein from a wide range of perspectives. Some of these studies have identified several critical governance challenges that have had tremendous impact on the sustainable exploitation of the various resources on the floodplain. Most of these challenges are closely related to the evolution of the institutional arrangements governing the exploitation of the resources stretching back to the colonial era. While the evolution of the governance institutional arrangements has been triggered by a combination of factors, government policies across the colonial, one-party and multiparty regimes have played quite a significant role. Policies in the areas of water, irrigation, land, and decentralisation have been quite influential. The impact of these policies on the resource management regimes in the floodplain has to a very large extent been shaped by the political context in which they were implemented.

This chapter does not seek to reinvent the wheel since an impressive array of studies on governance structures and challenges on the floodplain already exists. Using these studies and focusing on land and fishing, the main purpose of this chapter is to re-examine the governance challenges of the floodplain with the help of Helmke and Levitsky's (2004) typology of the patterns of interactions between formal and informal institutions as a basis for stimulating potentially new policy discourses about how the vast range of governance challenges can be reconceptualised and dealt with on a sustainable basis.

The Lake Chilwa floodplain is an important source of livelihood for people living on it and even beyond. It is estimated that the economic value of the wetland is about US$21 056 392 annually. The biggest contributor is fishing, which accounts for an estimated US$18 million, followed by agriculture at about US$1.5 million (Mvula and Haller 2009). The fisheries from Lake Chilwa and the wetland thus constitute an important subsistence and income-generating activity and contribute 20 to 23% of the total fish landing in Malawi. The Lake Chilwa fishery is important in terms of the number of people involved, the size of the catch and the number of auxiliary businesses that depend on the fishery.

The floodplain is famous for rice production, stream bank cultivation, and livestock grazing. In particular, the dry season provides excellent alluvial soil conditions for a second annual crop of rice or maize (Njaya 2009). The major crops cultivated in the area include rice, groundnuts, maize, tobacco, potatoes and sun flower. Land in the area is predominantly held under customary tenure whereby custody is vested in the traditional leaders. Formal irrigation schemes are located on leased public land whereas most of the estates are held in freehold (Peters 2004).

7.2 THEORETICAL FRAMEWORK: PROPERTY RIGHTS, INSTITUTIONS AND GOVERNANCE

The controversial debates about property rights are rooted in Hardin's (1968) classic work on the tragedy of the commons. The basis for the tragedy of the commons is that everybody's property is no one's responsibility to manage and preserve (Donda 1997). This situation leads to overexploitation or extinction of resources as each resource user tries to maximise their personal benefits from resource exploitation. The orthodox recommendation is that the resources should either be privatised or controlled by the state in order to prevent them from degenerating into victims of the tragedy of the commons (Ostrom 1990, Donda 1997, Hara *et al.* 2009).

Property rights are broadly defined as social institutions that define the range of privileges granted to an individual over specific assets such as parcels of land and water (Alchian and Demsetz 1973). This means that property rights entail mechanisms for defining and enforcing rights, social customs, legitimacy and recognition of the rights. For this reason, property rights are characterised as 'a bundle of characteristics [that inter alia] include exclusivity, transferability, inheritability and enforceability' (Hara *et al.* 2009). Thus property rights involve a variety of rights including the right to exclude non-owners from access, the right to appropriate rents and the right to sell or otherwise transfer resources to others (Alchian and Demsetz 1973).

Four different property rights regimes can be distinguished following Hardin's (1968) classic work on the tragedy of the commons: open access or free access (non-property); common property; state property; and private property. Notwithstanding the existence of the various forms of property rights regimes, the most critical point is that property rights perform the functions of limiting use, coordinating users and responding to changing resource conditions (Hara *et al.* 2009).

The question of property rights has been and continues to be a contentious one with regard to resource management of the nature and scale as that of the Lake Chilwa floodplain. The major challenge is that these resources are often treated as open access resources, as implied by Hardin. This perspective implies that the resources have to be managed; yet, often these resources had their own management regimes that regulated access, exploitation and preservation. Far from being open access resources, these resources were managed predominantly under common property rights regimes (Donda 1997, Hara *et al.* 2009), which worked because the pressures on the resources were manageable. The existing technology and the demand for the resources did not result in the overexploitation of the resources. In such a context, the management regimes are usually based on clearly delineated principles of membership, co-ordination, rules for harvest techniques, timing of use, knowledge sharing, and monitoring and sanctioning devices.

The quest to contract for property rights encapsulates processes of institutional change that have the potential either to promote efficiency or inefficiency in the exploitation of the resources. There is a constant search for appropriate property rights regimes because 'most management problems stem from incomplete, inconsistent, or unenforced property rights' (Hara *et al.* 2009: 523). As hinted above, property rights are a form of institution that, according to North (1990), are defined as rules of the game. Institutions are thus constituted by the rules and procedures that constrain some forms of behaviour and interaction between people and groups and enable others in social, economic and political domains. A distinction is often made between formal and informal institutions. The former are defined as rules and procedures that are created, communicated and enforced through channels widely accepted as formal; whilst the latter are defined as socially shared, usually unwritten, and created, communicated and enforced outside officially sanctioned channels (Helmke and Levitsky 2004).

Institutions are not static, a fact that is underlined by the pervasive quest for redefinition of property rights in order to enhance efficiency and effectiveness (Leftwich 2007). However, processes of institutional change are a deeply imbued political process that involves winners and losers. In this context, politics is understood as encompassing all activities of co-operation, conflict and negotiation involved in decisions about the use, production and distribution of resources, whether these are formal or informal, public or private, or a mixture of all (Harris 2006, Leftwich 2007). Thus politics is essentially about contestation and bargaining between interest groups with competing claims over rights and resources.

The processes of institutional change are, however, hardly neutral. They distribute advantage to some and disadvantage to others, expressing a mobilisation of bias in some way or another. Consequently, winners from the existing institutional arrangements fight for the maintenance of the status quo whereas losers vigorously push for changes in anticipation that such institutional reconfiguration would create opportunities for new streams of benefits (Leftwich 2007). In other words, processes of institutional change are often heavily contested by diverse interests with different forms and degrees of power, influence and authority, creating in the process winners and losers.

The efficacy of the resultant institutional arrangements is critically dependent on the interactions between formal and informal institutions. The patterns of interaction are a crucial determinant particularly in terms of whether any new institutional arrangements would promote pro-poor growth or not (Hare and Davis 2006). Helmke and Levitsky's (2004) typology of the probable patterns of interaction between formal and informal institutions is quite illustrative in this regard. As indicated above, there are four possible patterns of interaction, namely: complementary, substitutive, accommodating and competing (see Table 2).

Table 2: A typology of informal institutions

Outcomes	**Effective formal institutions**	**Ineffective formal institutions**
Convergent	Complementary	Substitutive
Divergent	Accommodating	Competing

Source: Helmke and Levitsky (2004)

The probable patterns of interaction are premised on two considerations: firstly, the degree to which formal and informal institutional outcomes converge or diverge; and secondly, the degree to which formal institutions are effective, that is, the extent to which institutions and procedures that exist on chapter are enforced and complied with.

Complementary informal institutions fill in gaps either by addressing contingencies not dealt with in the formal rules or by facilitating the pursuit of individual goals within the formal institutional framework. They also serve as a foundation for formal rules that might otherwise exist on paper, thereby playing a key role in making effective the formal rules of the game. *Accommodating* informal institutions create incentives for actors to behave in ways that alter substantive effects of formal rules but without directly violating them. They simply 'contradict the spirit but not the letter of formal rules' (Helmke and Levitsky, 2004: 729). Typically these informal institutions are often created by actors who dislike outcomes generated by formal rules but are unable to change or openly violate those rules. Even though accommodating institutions do not enhance efficiency, they are instrumental in promoting 'stability of formal institutions by dampening demands for change' (Helmke and Levitsky 2004).

Competing informal institutions structure incentives in ways that are incompatible with formal rules; in other words, to follow one rule, actors must violate another. This genre of informal institutions is often dominant in postcolonial contexts in which formal institutions were imposed on indigenous rules and authority structures. *Substitutive* informal institutions are employed by actors who seek outcomes compatible with formal rules and procedures but in a context where formal rules are not routinely enforced. In other words, these institutions seek to achieve what formal institutions were designed to do but failed to achieve. In the final analysis, the key question by Helmke and Levitsky (2004) is whether accommodating and substitutive institutions can contribute to the development of more effective formal structures or whether they crowd out such development by quelling demands for formal institutional change or creating new actors, skills and interests linked to the preservation of the informal institutions.

Helmke and Levitsky's framework provides some useful insights in thinking about processes of institutional change in a way that could potentially yield innovative governance regimes. Their framework highlights the importance of paying attention to the interaction between formal and informal institutions in order to fully understand the dynamics of institutional change. As Helmke and Levitsky (2004:728) observe: 'When institutions function effectively, we often assume that formal rules are driving actors' behaviour. Yet in some cases, underlying informal norms do much of the enabling and constraining that we attribute to the formal rules.' The dominant trend in natural resources management has been to supplant informal with formal management and governance regimes, based on the justification that traditional models of resource management are inefficient and predisposed towards overexploitation of resources (Donda 1997, Mvula and Haller 2009, Hara *et al.* 2009). However, empirical evidence suggests that wherever governments have attempted to formalise the management of natural resources, the resources have invariably degenerated into de facto open access status, leading to unsustainable exploitation and degradation of the resources.

7.3 GOVERNANCE CHALLENGES THROUGH THE HELMKE AND LEVITSKY FRAMEWORK

Despite the evolution in the property management regimes over the years, the Lake Chilwa floodplain is widely projected as a common pool resource. By definition, resources are difficult but not impossible to exclude from non-users and are subtractable (Mvula and Heller 2009). They are thus 'a class of resources for which exclusion is difficult and joint use involves subtractability of yield' (Hara *et al.* 2009:523). Scholars that have documented the evolution of the floodplain's management regimes of its resources suggest that the experiences have gone full circle. From exclusively a common pool property, the floodplain has degenerated into an open access resource underpinned by widespread pockets of informal privatisation. The consequence of these developments is that the floodplain now grapples with a wide range of problems, which include degradation, overtrapping and shooting of resident and migratory birds, population increase, soil erosion, destruction of breeding grounds and sanctuaries for fish, and the use of agro-chemicals that pollute the environment.

These developments are a culmination of complementary and contradictory streams of transformation spanning over four distinct periods of political, economic, social and even cultural history. The fate of the floodplain can broadly be charted over the precolonial, colonial, postcolonial, one-party and multiparty state dispensations (Njaya 2009, Mvula and Haller 2009). Some of the transformations are distinct to particular phases of history while others have endured between two or more phases with significant implications on the contemporary debates about the governance and management of the resources of the floodplain.

The forces that have precipitated changes in the resource management regimes have been predominantly exogenous, which, in turn, have spurred reactions and responses among the various local stakeholders within the framework of existing institutional arrangements (Peters 2004, Kambewa 2004, Njaya 2009). Our audit reveals a diverse range of stakeholders that have contributed to shaping the resource use patterns of the floodplain, including:

– informal local user groups who use land and wetlands and hunt birds;
– village leaders with powers to determine who can live in an area and access local resources and how much different resources different users are entitled to;
– district-level appointed officers enforcing government rules of natural resource and property ownership; and
– pressure groups lobbying for particular resource interests, notably birds.

The variety of stakeholders suggests that institutions and their links co-evolve as 'they are intertwined in ecologies of competition, co-operation and other forms of interaction' (Brown 2000:338). In other words, the environment adapts to institutions at the same time as institutions adapt to the environment. As a matter of fact, institutional histories require an understanding of both the origins of an institution and the paths along which it has developed.

Most scholars attribute the dramatic transformation in the management of the resources in the floodplain to the remarkable increase in their market values (Peters 2004, Kambewa 2005, Mvula and Haller 2009, Njaya 2009). The establishment of formal irrigation schemes in the area as early as the 1950s by both the colonial and postcolonial governments contributed to the increase in the value of the floodplain, making it more attractive for cultivation. The postcolonial government established four formal irrigation schemes as an integral part of the national irrigation schemes. The agricultural resources soon came under pressure as a result of the increased numbers of people seeking to access them, coupled with the natural population growth of the floodplain. Agriculture in the floodplain was commercially viable as a result of the increase in demand for food by urban and peri-urban consumers as well as buyers in the local markets (Peters 2004, Kambewa 2004). The establishment of the irrigation schemes was one of the significant pull factors since it revealed 'the attractions of [rice] as a food and cash crop' (Peters 2004:3).

Large numbers of people were attracted to the floodplain by a relatively vibrant fisheries industry in the wake of the economic crisis of the 1980s and 1990s (Njaya 2009, Mvula and Haller 2009). During this period, Malawi was implementing the economic structural adjustment programmes (SAPs) prescribed by the World Bank and International Monetary Fund as a means of kick-starting sustainable economic growth (Chipeta 1993, Chinsinga 2002). The economy was in a state of crisis because the public sector had not only overextended itself but also became grossly inefficient, corrupt and heavily statist orientated in its policies. As an integral part of the SAP policy prescriptions, a substantial number of people were retrenched in the public sector, estates and companies. This period coincided with the end of mine contracts in South Africa, which had absorbed a significant number of economically active male adults, who were now unemployed.

Most of these people were attracted to fishing in the then relatively pristine waters of Lake Chilwa. The commercial viability of the floodplain both in terms of the fishery and agricultural industries made it quite an attractive destination for unemployed and ambitious men and women. Moreover, all alternative water bodies had been overfished and the floodplain provided a lucrative alternative occupation for fishermen (Donda 1997, Mvula and Haller 2009, Njaya 2009). The migration of established fishermen from the somewhat depleted water bodies into the floodplain has had tremendous impact on the management strategies and regimes for the floodplain fisheries.

The pressure for agricultural land has continued to mount even in recent years. Access to wetlands for winter cultivation has become quite competitive following recurrent incidents of severe droughts. The intensification of the competition is also attributed to some recent government policy initiatives pertaining to irrigation, land and water resource management (Peters 2004, Kambewa 2004). It is further argued that wetlands tend to acquire the highest value following years of erratic and inconsistent rainfall patterns. Policy initiatives such as provision of treadle pumps, provision of inputs to farmers under the Targeted Input Programme (TIP), new land, irrigation, and water policies have heightened not only competition for access to wetlands but also changed the institutional dynamics of management and governance of these particular resources. There is, generally speaking, 'a growing conviction among government, donors and non-governmental organisations about the appropriateness of small-scale irrigation as a direction for development' (Peters 2004:6). Many farmers in the plain believe that wetlands or stream bank gardens 'constitute major sources of cash income to their families as well as contributing towards the family food supply' (*ibid*:6).

7.4 LAKE CHILWA FLOODPLAIN'S GOVERNANCE AND MANAGEMENT REGIMES

The major change in the management and governance regimes for the Lake Chilwa Floodplain resources was introduced in 1946 by the colonial government (Njaya 2009). Prior to 1946, the floodplain resources were governed under the common property rights regimes. According to Mvula and Haller (2009), in order to access resources, one had to belong to a clan by virtue of descent, marriage or even settlement as sanctioned by village heads of the respective localities on the floodplain. Migrants to a village or chiefdom were obliged to offer thanks, 'usually a small portion of the yield, to thank them [village heads] for allocating them a piece of land on the floodplain' (Mvula and Haller 2009:543). In vernacular, this is referred to as *chithokozo*, which literally means 'token of appreciation'. Similar practices are equally documented for the fishing industry. Each community claimed access to the beaches closest to them and membership of the community was quite an important determinant of access. There were several local institutions that governed access to exploitation and preservation of the fisheries. These were monitored and sanctioned by the local leaders. Evidence suggests that they worked well since 'they had traditional courts and also specific mystical powers to prevent misuse by threatening attacks by ancestral spirits' (Mvula and Haller 2009:540).

Besides granting chiefs and village headmen powers to allocate land and control the use of resources by their people and immigrants, the colonial administration departmentalised the management of the resources of the floodplain across the government's administrative machinery. Some of the key departments included the departments of water, wildlife, fisheries, and agriculture (Donda 1997, Njaya 2009). This period in history essentially marked the birth of the centralised mode of resource management, which was inherited by the postcolonial government up until 1995, just after the advent of a multiparty political system of government (Peters 2004, Njaya 2009). The fragmentation of resource management among the various departments made integrated management of the floodplain difficult. Consequently, the successful management of the floodplain resource would depend on an integrated rather than a departmentalised approach.

The major challenge of the centralised approach to managing the floodplain resources was that rules and guidelines were never enforced, neither for the fisheries nor for agricultural sectors. Thus while the state was legally available on the ground, it was institutionally absent to carry out enforcement, which promoted tendencies toward privatisation of the resources or the resources simply degenerated into an open access situation (Hara *et al.* 2009). There was thus basically an institutional vacuum that allowed stakeholders at the local level to experiment with all sorts of conceivable governance and institutional permutations. The floodplain essentially became a fertile ground for legal pluralism to thrive. This situation continues to be manifested in the contemporary debates about resource use and management in a bid to ensure sustainable exploitation.

The transition to democracy in May 1994 opened up space for a paradigm shift from centralised management to co-management as an integral part of the decentralisation policy reforms. The underlying philosophy of the reforms is to devolve power and authority from the centre to the periphery to ensure efficiency and effectiveness in the management of development programmes, including the management of natural resources. Broadly speaking, decentralisation is conceived as a critical element in facilitating the invigoration of the state institutional capacity in order to establish a credible foundation for an effective public sector but also as a means of bringing the state closer to the people (Chinsinga 2009). It is a means of dealing with the pervasive institutional weaknesses of the state, especially in the developing world. Proper decentralisation is associated with *devolution* of power, responsibility and authority, as opposed to *delegation*, which is a form of decentralisation that is mainly concerned with the transfer of resources to lower levels of government and not necessarily power, responsibility and authority.

The co-management paradigm has become fashionable in Malawi but it has only been fully implemented in the fisheries sector. According to Donda (1997), co-management entails partnership and sharing of power and responsibility between the government and resource users. The success of co-management is contingent on formal organisation, democracy, decentralisation of power and responsibility. The adoption of the co-management strategy in fisheries has led to the establishment of the beach village committees (BVCs) as representatives of resource users. According to Njaya (2009), as many as 48 BVCs have been established to regulate and oversee fishing activities in the floodplain.

In the agricultural sector, co-management is beginning to take shape under the auspices of irrigation. Under the new irrigation and water policies, communities that had their land appropriated from them in the 1970s are now having their land returned to them by government as irrigation schemes. Under these new policies, the government retains legal ownership of the schemes while their management is being handed back to communities (Zuka 2010). The government structures will only play a supervisory and monitoring role to help maintain standards. This development is, however, taking place at a time when the agricultural context has dramatically changed, especially in the wake of adverse climatic changes that have raised the value of wetlands. A number of policy initiatives are also well under way that may have tremendous implications for institutional arrangements on the ground.

Two issues are worth raising at this point. First, Malawi's new land policy designates wetlands as unallocated land and, second, its new water policy proposes licensing all categories of water users for them to access water (Peters 2004, Chinsinga 2008). The land policy suggests that wetlands are an open access resource yet empirical evidence clearly shows that there are no idle wetlands. They are owned under an ensemble of property rights regimes that have evolved in response to the changing context in the Lake Chilwa floodplain. The water policy's proposal to license water users across all categories raises more questions than answers given the configuration of current water users. The main competitors in water use are stream bank gardens, wetland gardens, irrigation schemes, private estates and households (Peters 2004, Kambewa 2004). The main challenge is that the prescriptions of the key policy documents are not based on full understanding of the existing patterns of resource use and exploitation, which complicates the interface between formal and informal institutions. The policy documents are often conceived on the assumption that there is a vacuum as far as the management of the resources is concerned, yet there are 'complex ways in which use of watered lands by different groups interact, which has significant implications for co-ordinated action across sectors' (Peters 2004:8).

7.5 APPLICABILITY OF HELMKE AND LEVITSKY'S FRAMEWORK TO THE FLOODPLAIN

As I argued earlier, Helmke and Levitsky's (2004) framework is not applicable to understanding the dynamics and patterns of resource access, use, and management in the pre-colonial era. As observed above, the management and governance of the resources were effectively and efficiently mediated through informal institutions (Donda 1997, Njaya 2009). These institutions were effective and efficient in promoting not only equitable but also sustainable exploitation of the resources in many ways by 'the spiritual affinity with the natural environment and traditional customs' (Donda 1997:22). The governance structures had significant purchase on resource use and exploitation because the regulations reflected, or were at least in tune with, widely shared notions of social justice. In other words, compliance with rules of resource use and exploitation was generated by the socio-cultural pressures that basically defined what was acceptable or not.

Now the situation is entirely different from that in the colonial era. The pre-colonial modalities of resource access, use and management were sidelined by the introduction of a departmentalised approach to resource governance and management in the floodplain (Mvula and Haller 2009). This institutional innovation in the management of Lake Chilwa Floodplain brings the Helmke and Levitsky (2004) framework to bear on the resultant patterns and dynamics of resource access, use and management. The main reason for this is that the departmentally inspired guidelines for resource governance and management were never effectively enforced. The paradox is that the state had assumed the mandate for resource governance and management but it was institutionally absent on the ground. According to Mvula and Haller (2009), prior to 1996 no enforcement was carried out as part of departmental activities. For instance, the Department of National Parks and Wildlife only established its presence in the floodplain in 1997 when it was declared a Ramsar Site. The capacity of the Department of Fisheries has always been limited due to budgetary constraints and, in addition, formal local committees with interest in the management of the resources in the floodplain were non-existent.

The ineffectiveness of the colonial and post-colonial state management resource regimes in the floodplain combined with the weakening of local institutions such as traditional values, norms and beliefs, chieftaincy and local resource user groups created conditions of open access that culminated in the rampant degradation of the resources in the floodplain. The vacuum created a favourable atmosphere for the substitutive interface between formal and informal institutions. The implication of this form of interface is that informal institutions supplant formal institutions because they are extremely weak. Informal institutions that emerge intend to achieve the goals that formal institutions intend but are unable to achieve simply because they are weak (Helmke and Levitsky 2004). As demonstrated above, the main challenge is that the institutional alternatives that emerged as a result of the vacuum have not been fully documented and studied for lessons to feed into the current efforts to propagate viable management regimes of the resources. The oversights in the water and land policies clearly manifest a lack of appreciation for how essentially substitutive informal institutions can be exploited to creatively develop robust resource management regimes.

The evidence shows that the post 1995 period, characterised by a paradigm shift from centralised state control of resources to co-management, has created a favourable environment for the competitive interface between formal and informal institutions. Unlike in the substitutive interface, the outcomes intended by formal and informal institutions are divergent (Helmke and Levitsky 2004). This presents considerable challenges for the efforts to come up with viable resource management regimes since the underlying logic between the two sets of institutions are simply irreconcilable.

Irreconcilable challenges are apparent in both the fisheries and agricultural sectors. In the fisheries sector, the BVCs have assumed centre stage. They were established under the auspices of co-management as a means of ensuring the participation of local communities (Donda 1997, Njaya 2009). However, the BVCs have not been as effective as they had been envisaged. The conclusion of various studies is that the establishment of the BVCs did not fully take into account or build on a vast array of informal institutions that were instrumental in regulating the fisheries sector during the period of institutional void following the entrenchment of state regulation in the exploitation of the resources in colonial and postcolonial eras. There are, for example, enduring conflicts between BVCs and local leaders that have greatly undermined the overall efficacy of the BVCs (Donda 2003, Njaya 2009, Mvula and Haller 2009).

As legal rational entities, BVCs have to assert themselves as regulatory bodies for fisheries at the local level. This has, however, brought them on a collision course with traditional leaders as the BVCs have sought to challenge them on the *mawe* practice. Traditional leaders are beneficiaries of an old age custom of *mawe*, a form of tribute that fishermen pay to them for plying their trade in the waters under their respective jurisdictions. The BVCs are determined to stamp out the *mawe* practice. However, traditional leaders are defiant. They use their powers to allow fishermen to operate illegally during closed seasons because they are interested in the *mawe*. Since traditional leaders wield considerable influence, the authority of BVCs is undermined, leading more or less to the failure of the co-management scheme. The conflict is inevitable because of the way in which BVCs are constituted. They are facilitated by officials from district headquarters. Traditional leaders do not have a say with regard to who sits on BVCs but are co-opted as ex-officio members. Their role is to provide advice but in practice the BVC members have tended to sideline them altogether.

The activities of BVCs are further undermined by the unfinished business of decentralisation (Mvula and Haller 2009). There is considerable ambiguity in the actual roles of BVCs especially since they are not formally recognised as an integral part of the decentralisation institutional framework. According to Njaya (2009), there is limited support for BVCs from the district councils, which would be critical for the BVCs to fully assert their authority. BVCs further struggle for power with other local level institutions such as the village development committees (VDCs). Traditional leaders take advantage of the fact that they chair VDCs to further marginalise BVCs, which they regard as a threat to rent seeking opportunities including *mawe*. The conflict between VDCs and BVCs is inevitable because the remit of the former is broadly defined as development. Fishery is regarded as an integral part of development and since the BVCs want to stamp their authority in this area, the traditional leaders use their weight as chairpersons of VDCs to completely overshadow BVCs.

In the agricultural sector, the competitive interface between formal and informal institutions revolves around the issues of land allocation and water use (Peters 2004). The essence of the situation is that the non-enforcement of the departmentalised policy guidelines has promoted informal privatisation or open access to agricultural resources in the floodplain. Non-enforcement of the rules and regulations has further allowed traditional leaders to entrench themselves as gatekeepers who regulate access to agricultural resources in the floodplain. *Chithokozo* has now evolved into a monetised and extractive practice.

7.6 CONCLUSION

In this chapter, I re-examined the governance challenges of the Lake Chilwa Floodplain, using Helmke and Levitsky's (2004) framework. The main premise of this framework is that understanding the various patterns of interaction between formal and informal institutions could be an extremely useful basis for stimulating new policy discourses about natural resource governance and management challenges in general and with particular reference to the Lake Chilwa floodplain. In re-examining these governance challenges, my ultimate objective was to provoke a new research agenda that would make use of the framework to shed light on various dimensions of the governance challenges in access, use and management of the Lake Chilwa floodplain resources.

From the evidence, I argue that the effective management and governance of natural resources is not only about putting in place formal institutions. It is also about understanding the whole institutional landscape both in a historical and contemporary sense in order to establish a credible basis for institutional designs that would better deal with various forms of governance challenges. This means that the interface between formal and informal institutions in designing governance and management regimes is critical. The key message here is that addressing formal institutions is often not the 'magic bullet' solution since sometimes informal institutions are critical to the effective and efficient functioning of formal institutions. Some of these informal institutions include traditional values, beliefs and practices and social organisations. Further analysis is required that pays attention to the various patterns of interface between formal and informal institutions to create viable, if not sustainable, governance and management regimes for valuable resources in the Lake Chilwa basin.

REFERENCES

Alchian A and Demsetz H. 1973. 'The Property rights paradigm', *Journal of Economic History* 33: 16-27.

Brown B. 2000. 'Do institutions matter? ' pp. 331-339, in: *Comparative Politics: Notes and Readings,* 9th Edition. Harcourt, UK.

Chinsinga B. 2002. 'The politics of poverty alleviation in Malawi: A critical review', pp. 25-42, in: Englund H (ed) *A Democracy of Chameleons: Politics and Culture in the New Malawi.* Uppsala, Nordiska Africa Institute, Sweden.

Chinsinga B. 2008. 'Exploring the politics of land reforms in Malawi: A case study of the Community Bases Rural Land Development Project (CBRLDP)', *IPPG Discussion Paper Series No. 20.* University of Manchester, United Kingdom.

Chinsinga B. 2009. 'Politics of poverty reduction in Malawi: A local governance perspective', in Matlosa K (ed) *The State, Democracy and Poverty Eradication in Africa.* Electoral Institute of Southern Africa (EISA), Pretoria.

Chipeta C. 1993. *The Impact of Structural Adjustment on the People of Malawi: The Impact of Structural Adjustment on the Population of Africa.* James Currey, London.

Donda S. 1997. 'Fisheries co-management in Africa: Case study of Lake Chiuta', pp. 21-40, in: Nielsen A and Sverdruo-Jensen J (eds) *Fisheries Co-management in Africa: Proceedings from a Regional Workshop on Fisheries Co-management Research, August 1997.* Department of Fisheries, Mangochi, Malawi.

Hara M, Stephen T and Matose F. 2009. 'Governance of the commons in Southern Africa: Knowledge, political economy and power', *Development Southern Africa* 26(4): 521-537.

Hardin G. 1968. 'The tragedy of the commons', *Science* 162: 1243-1248. Available at: www.sciencemag.org/cgi/reprint/162/3859/1243.pdf (Accessed on 10th April 2010).

Hare P and Davis J. 2006. 'Institutions and development: What we (think we) know: What we would like to know.' *IPPG Discussion Paper Series No. 3.* University of Manchester, United Kingdom.

Harris J. 2006. 'Notes on historical institutional approach to the IPPG Agenda.' *IPPG Discussion Paper Series No. 1.* University Manchester, United Kingdom.

Helmke G and Levitsky S. 2004. 'Informal institutions and comparative politics: A research agenda', *Perspectives on Politics* 2(4): 25-740.

Kalk M, Mclachland A and Williams H. 1979. *Lake Chilwa: Studies of Change in a Tropical System.* Dr. W. Junk Publications: The Hague, Netherlands.

Kambewa D. 2004. *Patterns of Access and Use in Wetlands: The Lake Chilwa Basin.* Bunda College of Agriculture, Lilongwe, Malawi.

Kambewa D. 2005. 'Access to and monopoly over wetlands in Malawi.' *Paper presented at the International Workshop on African Water Laws: Plural Legislative Frameworks for Rural Water Management in Africa, Johannesburg, South Africa. 26−−28 January 2005.*

Leftwich A. 2007. 'The political approach to institutional formation, maintenance and change: A literature review essay.' *IPPG Discussion Paper Series No. 14.* University of Manchester, United Kingdom.

Mvula P and Haller T. 2009. 'Common pool resource management in Lake Chilwa, Malawi: A wetland under pressure', *Development Southern Africa* 26(4): 539-553.

Njaya F. 2009. 'Governance of Lake Chilwa common pool resources: Evolution and conflicts', *Development Southern Africa* 26(4): 663-676.

North D. 1990. *Institutions, Institutional Change and Economic Performance.* Cambridge University Press, Cambridge.

Ostrom E. 1990. *Governing the Commons: Evolution of Institutions for Collective Action.* Cambridge University Press, Cambridge.

Peters P. 2004. *Informal Irrigation in Lake Chilwa Basin: Stream Bank and Wetlands Gardens.* Kenny School of Government, Harvard University, US.

Zuka S. 2010. 'Opportunities and constraints in smallholder irrigation management schemes: The case of *domas.*' *Unpublished MA thesis in Development Studies.* Chancellor College, Faculty of Social Science, Zomba Malawi.

Chapter 8

THE ECONOMY AND LIVELIHOODS OF THE LAKE CHILWA BASIN

Milton KUTENGULE

8.1 INTRODUCTION

In this chapter, I will analyse the economy and livelihoods of the people of the Lake Chilwa plain in the south-eastern part of Malawi, with a view to identifying the interactions between the ecosystems of the area and the exploitation of its natural resources against the background of recessions in the water levels of the lake. In the process, I identify gaps in the analysis and policy appreciation of important links between local economic activities and national level policies and programmes of action.

The literature on the links between the performance of the Malawian economy and livelihoods of ordinary Malawians, especially the poor, in the post-independence era (the period after 1964) was remarkably small until the 1990s. This was due to government's pre-occupation with economic growth measures, which had little consideration for the active participation of the poor in the economy. The limited important works up to the 1980s are Humphrey (1973), Gupta (1974), Acharya (1978), Kydd and Christiansen (1982) and the Malawi Government Annual Economic Reports from 1970 to 1990. However, from 1990 onwards, the literature on Malawi's economic growth prospects with a concern for the poor has grown considerably owing to the realisation that the economic structural adjustment policies (SAPs) that the International Monetary Fund and the World Bank had prescribed for Malawi and other African countries in the 1980s had not stimulated the expected high rates of economic growth and poverty reduction (World Bank 1989, Pryor 1990, World Bank 1990 a and b, Government of Malawi/United Nations 1993, Cornia and Helleiner 1994).

Some studies have attributed the failure of SAPs to a lack of understanding of the links between the policies, markets and infrastructure at the meso (middle) level and micro (especially household) levels (Cornia *et al.* 1987, World Bank 1989, World Bank 1990a and 1990b). This resulted in the adoption of analytical frameworks that integrated household and enterprise dynamics in policy simulations such as the Meade-Salter-Swan models, computerised general equilibrium (CGE) models and social accounting matrices (SAMs) (World Bank 1990b). These frameworks emphasise the need to disaggregate policy analysis to identify the distributional effects of macroeconomic policy on various socio-economic groups and parts or regions of a country. However, micro and meso levels analyses have themselves been hampered by lack of reliable data on the links between the micro-economy and the macro-economy. The data gap has made it difficult to understand how national level policies have impacted on local economies (such as that of the Lake Chilwa plain) and the livelihoods of people of such economies.

The paucity of literature, information and data on the economy and livelihoods of the Lake Chilwa plain is profound, as most studies on the area tend to focus on the Lake Chilwa fishery largely from the environmental, rather than economic, angle. Due to a desire to explore the relationship between environmental and economic processes in the area from the late 1990s, there has been an increase in studies that have enumerated economic activities but are based on the Lake Chilwa resources instead of the whole Lake Chilwa plain as a coherent unit of economic analysis or a local economy (see, for example, World Food Programme *et al.* 1996, Schuijt 1999, Overseas Development Group 2002a and b, Peters 2004, Kalindekafe 2010). To my knowledge, no study has so far marshalled the empirical evidence for assessing the economic contribution of the Lake Chilwa plain (including the Lake Chilwa fishery) to the national economy. In this chapter, I therefore attempt at a systematic review of the empirical evidence on the nature and size of the economy so as to identify potentially significant sources of growth in the economy, constraints to its growth, livelihood opportunities, and the sustainability of these livelihoods against a background of a volatile ecological system and climate change. I also explore differences in livelihood experience across localities within the economy and problems of estimating sectoral contributions to the growth of the Lake Chilwa plain economy.

Based on the analysis of the major economic activities of the floodplain inhabitants, I identify areas of further action research to support the management of the common property resources of the Lake Chilwa plain, as well as possible policy concerns and how they could be addressed in order to promote the sustainability of the natural resource base and livelihoods. I draw on the findings of various research projects undertaken in the area, especially the Livelihoods and Diversification Directions Explored by Research (LADDER) Project conducted by the University of East Anglia in two villages located on the western shore of Lake Chilwa, particularly in the Zomba District, in the early 2000s. The other important information I analyse is gleaned from the studies or surveys conducted in the Lake Chilwa plain by the Malawi Fisheries Department, the National Statistical Office (NSO), University of Malawi (Chancellor College) researchers, and the Broadening Access and Strengthening Input Market Systems- Collaborative Research Support Programme (BASIS-CRSP) in the early 2000s, which was funded by the United States Agency for International Development (USAID). I also draw from my own research and other experiences in the same plain to analyse socio-economic phenomena in the area.

8.2 THE LAKE CHILWA PLAIN

The Lake Chilwa floodplain is the catchment area of Lake Chilwa, the second largest lake in Malawi after Lake Malawi. Lake Chilwa is a shallow inland drainage (without any outlet) that does not exceed 6 m depth at peak water levels. This shallowness results in an efficient nutrient recycling system such that the lake is one of the most productive in Africa (Chiotha 1996).The open water area of Lake Chilwa is typically 678 km^2 and is surrounded by 600 km^2 of swamps, 390 km^2 of marshes and 580 km^2 of seasonally inundated grassland on the floodplain. Any of these vegetation types varies with the level of the lake within a year and inter annually. Much of lake's eastern shore lies in Mozambican territory and there are cross-border movements of people, goods and services between Malawi and Mozambique across the lake. However, there is little information from reliable studies on the economy and livelihood systems on the Mozambican side of the lake. Almost all residents of the floodplain originated in Mozambique and the Nyanja tribe is the oldest in the area (Vaughan 1982, White 1987).

The Lake Chilwa plain is located in south-eastern Malawi, in the districts of Machinga, Zomba and Phalombe, along the western shore of the lake. It is located in the area between Mlomba, Mbonechela and Ntaja to the north of the lake; the Chikala Hills and Zomba Plateau to the west; the Buleya and Ntonya areas to the south-west; Migowi and Phalombe towns to the south; and the Mpoto Lagoon to the south-east of the lake (Kalindekafe 2010).

The Lake Chilwa plain is a highly variable ecosystem with seasonal and periodic water level fluctuations that have important implications for variability in the livelihood activities of its residents. The catchment area on the Malawi side is 5 669 km^2 while that on the Mozambique side is 2 680 km^2. The Malawi 2008 population and housing census estimated that the three districts that surround Lake Chilwa (Machinga, Zomba and Phalombe) had a population of 1 471 661, which had risen by 28% from the 1998 population of 1 148 265. Such a high rate of increase may be attributable to a high fertility rate and high levels of migration into the area from other parts of the country and Mozambique. The two censuses found that the majority of the inhabitants of the plain were from the Nyanja, Lomwe, and Yao tribes (NSO 1998 and 2008, Peters 2004).

8.3 THE ECONOMY AND LIVELIHOODS OF THE LAKE CHILWA PLAIN

The nature or structure, size and performance of an economy, including that of Malawi, is conventionally conceptualised and analysed in terms of changes in its total production (capacity) across the various sectors of the economy. This has been the tradition in Malawi where annual economic reports have consistently been prepared for more than three decades. This way of analysing the economy is known as the production approach which is based on the system of national accounts (SNA) originally developed by the United Nations in 1968 but revised jointly by the United Nations itself, the European Union, the Organisation for Economic Co-operation and Development, the World Bank, and the International Monetary Fund in 1993 (Barth and Hemphill 2000). The system was developed as an accounting framework within which macroeconomic data such as total output, rates of inflation and unemployment, exchange rates, savings and investments, and balance of payments can be compiled and analysed. The system is used for reporting national accounts data that conform to internationally accepted definitions and classifications of economic activity thus allowing for comparisons among economies. For the Lake Chilwa plain to be considered an economic entity there is a need to adopt the same conceptual and analytical framework as the national one so that the sectoral and economy-wide constraints to growth and livelihood composition or promotion can be in harmony with the national economic and development agenda.

Total production (or gross output) of an economy is the value of all goods and services produced in the economy minus total production costs and depreciation over a specific period of time (usually per annum) that is sold to consumers, other enterprises and foreigners. Production costs and depreciation are subtracted from total production so as to focus on changes in production or net income, which is called the value added. Non-market or subsistence output is also added to the value added because it can, in some cases, quantitatively exceed measured formal sector production as is the case in Malawi. The total production derived is called the gross domestic product (GDP) which is the sum of value added across all producers or sectors in the economy. This explains why the economy of the Lake Chilwa plain cannot meaningfully be confined to the analysis of extraction of value from the Lake Chilwa resource alone as some recent analyses have done. There is a need to cover all the sectors of the floodplain, following the conventional approach.

To illustrate the discussion above, based on the production approach, the GDP of the Malawi economy is measured as the total of value-added derived from all sectors of the economy and its typical structure is as presented in Table 3 below.

Table 3: Malawi's GDP by sector of origin at 2007 constant prices, 2008, 2009 and 2010

Sectors	2008	2009	2010
	Millions of Malawi kwacha		
Agriculture, forestry and fishing	154 157	174 304	177 752
Mining and quarrying	7 883	8 268	14 903
Manufacturing	63 328	66 347	69 169

Sectors	2008	2009	2010
	Millions of Malawi kwacha		
Electricity, gas and water	8 717	9 289	9 596
Construction	16 475	17 697	20 621
Wholesale and retail trade	118 742	126 584	133 985
Transport and storage	21 697	23 629	24 614
Accommodation and food	11 123	12 595	13 632
Information and communication	22 087	24 409	26 652
Financial and insurance services	25 979	28 003	30 983
Real estate activities	29 592	33 178	36 834
Public administration and defence	13 953	14 637	15 488
Total GDP at 2007 constant prices	553 100	602 567	642 816

Source: Government of Malawi (2011)

Table 3 above gives a summary of the structure of the Malawi economy and its major sectors, which were Agriculture, Distribution (wholesale and retail), Manufacturing, Real Estate Services, and Financial and Insurance Services, in descending order of quantitative significance, between 2008 and 2010. The size of the economy of the Lake Chilwa plain has never been estimated systematically and coherently to provide information on its annual total income, as for one economic entity. Since its total value-added has never been computed across the sectors, it is difficult to present it in the format of Table 1 above. However, there are fragments of data on some of the sectors reflected in the table but not necessarily for the floodplain as one economic entity. From the structure presented in Table 1, it is evident that the economy of the Lake Chilwa plain cannot be limited to the fisheries sector which is only subsumed as part of the agricultural sector in the national accounts. Subsuming the fisheries under the agriculture sector has led to the understatement of the significance of the fishing industry in the national economy.

Despite the absence of economy-wide analysis for the floodplain, there are fragments of socio-economic analyses conducted in various parts of the plain at different points in time and this limits the extent to which vital socio-economic processes can be generalised at the Lake Chilwa floodplain level. Nevertheless, some surveys or studies have documented some of the major economic activities undertaken in the area: agricultural production, fish production, trading in fish and non-fish items, beer-brewing, employment in various sectors, repair services, provision of personal services such as hairdressing, and providing transport services, especially using bicycles (locally known as *kabaza* or *dampa*), which operate from some points along the main tarmac road between Blantyre and Liwonde and major trading centres along the western shore of the lake. (Overseas Development Group 2002a, World Food Programme *et al.* 1996, Malawi Government 1994). My own field observations confirm the dominance of these economic activities in the area. Non-farm economic activity is concentrated along the tarmac road and at the trading centres that have electric power supply from the national grid operated by the Electricity Supply Commission of Malawi (ESCOM). The main trading centres are Migowi in Phalombe, Matiya, Govala, Bimbi, Namasalima, Songani, Domasi, Namwera Turn-Off in the Zomba District; and Mpheta, Mpita, Mposa and Nayuchi in the Machinga District. The availability of power in these trading centres has stimulated local economic activity as it has provided various business opportunities that were not feasible when there was no electricity in these places. However, there is lack of data on the volume and impacts of these economic activities and the major factors influencing their success or otherwise.

In the above sections, I have analysed the data and information on the major economic activities gleaned from various surveys or studies conducted in the plain. I have drawn from my own research and other experiences in the same area to analyse the links between economic and environmental processes in the plain and how they impact on livelihoods in the area.

8.3.1 Agricultural production

The Lake Chilwa plain is a highly productive agro-ecological zone as it has rich sedimentary alluvial soils. Agricultural production in the area is dominated by small-holder agriculture with landholding sizes averaging mostly less than one hectare per household (Malawi Government Undated). The National Sample Survey of Agriculture (NSSA) conducted by the National Statistical Office in 1992/1993 found that estate or commercial farming was not significant within the plain and most of the estates grew tobacco in the Zomba and Machinga Districts. These tobacco estates provide some employment opportunities to unskilled workers. The other major crops that are grown in the plain are rice, maize, millet, sweet potatoes, cassava, pigeon peas and beans. Recently, burley tobacco has been introduced in the area as a cash crop that is grown in almost all parts of the Lake Chilwa plain. Within the plain, there are several rice production schemes introduced by the governments of Malawi and the Republic of China (Taiwan) in the 1960s and the 1970s. The major rice schemes are the Domasi Irrigation Scheme in the Machinga District, and the Likangala, Khanda and Njala Schemes in the

Zomba District. Farmers at these schemes, as is the case with the tobacco estates, provide casual wage or piece work employment (known locally as *ganyu*) during much of the year thus addressing food security problems during the hungry season.

Recently, these schemes have enabled farmers to grow maize, rice and water melons during the dry season using irrigation. This dry season cropping has significantly increased the income, employment opportunities and food security of the people of the area as a result of increased crop production. The harvest of these crops coincides with the onset of the hungry season in November and December when households run out of food stocks from the rain-fed crop production harvested between April and August. However, due to a limited water supply during the dry season, the number of farmers who can grow these dry-season crops on the schemes is restricted. Even those that are allowed to grow these crops struggle to get sufficient water for their dry season crops (Zuka 2009). Peters (2004) also documents cases of upstream-downstream competition for irrigation water in times of water scarcity between stream bank gardens, wetland gardens, irrigation schemes, and private estates in various parts of the Lake Chilwa plain. This underlines the fact that, with increased irrigation capacity of the schemes and other parts of the Lake Chilwa plain, the area can significantly increase its crop production, thereby contributing to the national economy and food security. In the largest of the irrigation schemes in the floodplain, the Domasi Irrigation Scheme, which has about 500 hectares, farmers earn an average of about US$555 per farmer per annum which is much higher than the estimated average of US$277 for the whole floodplain (Kalindekafe 2010). From these figures, it is clear that access to irrigation services enables farmers to increase their agricultural production substantially as they can grow more than one crop a year.

Estimates of total agricultural production are not consolidated at the Lake Chilwa plain level but some data is generated by the National Crop Estimates Committee through the Ministry of Agriculture at some Extension Planning Areas (EPAs) that fall within the Lake Chilwa plain. The EPAs that border the lake are Nampeya, Nanyumbu and Nsanama in the Machinga District, Ntubwi, Nsondole and Mpokwe in the Zomba District, and Tamani and Mpinda in the Phalombe District (World Food Programme *et al.* 1996). Estimates of agricultural production in these EPAs give a picture of crop production. While these crop production estimates give a rough picture of the nature and magnitude of agricultural production, their accuracy is a subject of debate on issues of methodology and the quality and motivation of data collection staff.

However, the data can form a starting point in estimating the level of agricultural production in the area, which could not be done for this chapter due to time and other resource limitations. It is noteworthy, nevertheless, that the LADDER Project (Overseas Development Group 2002a and b) and the World Food Programme *et al.* (1996) found that crop yields had been declining from the 1990s due to erratic rainfall which had a range of negative impacts: flooding and droughts; pest attacks; declining soil fertility due to lack of crop rotation in the face of rapid population growth; labour constraints; rising costs of farm inputs; inefficient input and produce marketing largely due to poor road infrastructure; and a lack of agricultural extension and credit services. The implication here is that addressing these constraints would foster agricultural production in the plain, thereby boosting its contribution to the national economy, food security and socio-economic well-being at both local and national levels. Since 2004, however, maize production and food security have improved in the plain due to a combination of good rainfall and greater availability of subsidised fertilisers and seed. The other constraints will still need some consideration in order to raise agricultural productivity and production.

In addition to crop production, livestock are kept by the people of the plain, especially chickens, ducks, goats and cattle (Malawi Government Undated; Kutengule 2000, Overseas Development Group 2002a and b). Various payments or user charges were introduced on services that were previously free of charge and this has increased livestock mortality by making animals more vulnerable to disease. Livestock production has been negatively affected by a lack of veterinary assistants, the closure and partial privatisation of dip tanks and veterinary services, cattle rustling, diminishing grazing land and the high incidence of livestock disease.

8.3.2 Fisheries

Fisheries production within the lake has a particularly high potential to contribute to local and national economic growth and poverty reduction as Lake Chilwa is one of the most productive lakes in Africa. However, most recent socio-economic studies focusing on poverty issues in Malawi exclude fisheries-based livelihood systems in their analyses. Consequently, national perceptions on poverty reduction and development initiatives lack a clear and satisfactory understanding of the links between economic growth processes and the fishing economy. Over 60% of animal protein in the Malawian diet is derived from fish and its by-products or derivatives (Malawi Government 2006). Furthermore, the fisheries sector contributes directly to the incomes of about 300 000 people, excluding fish traders whose total estimate is not known. Nevertheless, it is believed that nearly 1.6 million people in lake shore communities derive significant proportions of their in-kind and cash incomes from the fishing industry. The number of fish traders would include people at various levels from artisanal (small-scale) to industrial operators. However, in terms of national accounts, fisheries contribute a relatively small proportion (4%) of the gross domestic product, although its importance in local economies cannot be overemphasised (Malawi Government 2006).

Climate change processes contribute to fluctuations in the levels of Lake Chilwa, which affect the productivity of the lake's fishery as well as the livelihoods of the people of the plain. According to Lancaster (1979), the lake dried up in 1900, between 1914 and 1915, in 1922, between 1931 and 1932, in 1934, in 1954, between 1960 and 1961, in 1967, and in 1973 due to a combination of dry spells and high levels of evaporation. When the lake level declines significantly, as it did recently in the dry years of 1967 and 1995, fish catches dwindle drastically with serious effects on the economy of the floodplain. However, following such episodes of drying up, the fishery of the lake usually recovers fully after three to four years. This relatively quick recovery is aided by fish residues in the deltas of the Likwenu, Domasi, Songani, Likangala, Namadzi, Thondwe, and Phalombe rivers and the Mpoto Lagoon on the southern end of the lake in the Phalombe District.

The declines in the level of the lake and fish catch coincide with poor agricultural production and hunger occasioned by poor rainfall and droughts. These sources of re-population of the lake after refilling with rainfall are protected through the banning of certain unsustainable fishing methods especially the use of seines and poisonous plants. After the lake's recession of 1995, the government came up with measures, especially the establishment of Beach Village Committees, so as to work with fishing communities to facilitate increases in the fish stock. Unfortunately such measures have tended to create conflicts between government officials, local leaders and fishers over control of the fishery (Overseas Development Group 2002a and b).

According to Njaya (2002), fishing activity on Lake Chilwa increased tremendously between the mid-1970s and the late 1990s as evidenced by increases in fishing gear, craft, number of gear owners, crew members, boats, traps and in-migrating fishers from Lake Malombe and Lake Chiuta. Similarly, the LADDER Project has documented population movements from one part of the western shore of the lake to another principally in search of farming and fishing opportunities. These movements were mostly from the densely populated Phalombe District to the Zomba part of the western shore as well as from Mozambique into Malawi (Overseas Development Group 2002b). The increase in the fishing effort, a rise in population movements, increasing population density on farm land and the closure of employment opportunities on South African mines have combined to exert pressure on the fishery of Lake Chilwa with negative effects on the livelihoods of the people of the area. The LADDER Project has documented these effects as: diminishing agricultural productivity and production due to increased pressure on the land; diminishing fish catch and earnings from fishing; conflicts between indigenous residents and immigrants over fishing rights, farm land, and grazing land; an increase in bribery and corruption over access to fisheries, land and other natural resources; and a rise in the incidence of infectious diseases, especially HIV and AIDS (*ibid*).

In the 1960s, Malawi's annual fish production was estimated to have varied between 11 600 and 41 700 metric tons. After the 1970s, however, national fish production declined from a peak of 81 800 metric tons (in 1987) to the lowest annual catch of 33 895 metric tons in 1995. This is the year when Lake Chilwa dried up and its annual catch was only 1 328 metric tons, which contributed only 4% of total national fish production in that year. This decline in national fish production, a result of the drying up of Lake Chilwa, underlines the strategic importance of the lake in national fish production and economic and nutritional well-being. The fish is consumed locally as well as exported especially to Zambia and Zimbabwe largely through informal cross-border transactions (Salama and Jones 1982).

In the recent past, with domestic fish production declining, fish imports in Malawi have taken a generally upward trend from about 760 metric tons in 1997 to 2 808 metric tons in 1999 and 1 630 metric tons in 2000. Concurrent with this increase in fish imports was a decline in fish exports, indicating that the fisheries sector has not contributed significantly to Malawi's foreign exchange earnings although it has the potential to do so. The average fish catch on the lake is 13 000 metric tons but in some years the catch rises to about 25 000 tons and in peak fish production years, the lake contributes up to 43% of Malawi's total annual catch. However, much of the catch is widely sold in domestic markets, especially in Southern and Central Malawi. Less than 10% of fresh catch is supplied to the local markets of Mposa, Mpheta, Nkuba, Kachulu, Zomba City, Domasi, Songani, Migowi and Sombani. The fishing industry around the lake also provides employment opportunities to a significant number of people in fishing, fish processing and fish trading thus supporting the livelihoods of both rural and urban families, totalling about 40 000 people. Ancillary activities are undertaken by at least 200 000 additional people (Malawi Government 2000).

Fishing, fish processing and trading contribute to livelihood systems in the forms of incomes (cash and in-kind) and employment opportunities especially in areas that surround fisheries. As noted in Chapter 2, the Lake Chilwa fishery contributes significantly to total national fish production although its contribution to the national economy has not been analysed adequately. The fishery is thus a vital source of livelihood for people who live around it on both the Malawian and Mozambican sides and the fisheries of the plain have an even greater potential to contribute to the national economy because the plain has other fisheries (in addition to Lake Chilwa) in the form of the Mpoto Lagoon and many rivers that flow into the lake. In addition to fisher folk and communities in fishing areas, there are other people who benefit directly from upstream and downstream fishing activities and associated activities. Fishing is also undertaken as a livelihood strategy either seasonally (when catch is abundant) or on a permanent basis (Bene *et al.* 2007) but forms an important part of households' livelihood strategies. In some cases, such as the Zambezi Basin, fishing contributes more to household incomes than do cattle and crop sales (Turpie *et al.* 1999). Similarly the LADDER Project estimated in the early 2000s that incomes from fishing in the Lake Chilwa villages of Katanda and Sauka-Phimbi in Zomba contributed more income than the other sources (Overseas Development Group 2002b).

From the above studies, it is evident that fishing is a central element of the livelihood systems of the Lake Chilwa plain. It is part of a flexible set of economic activities that constitute diversified livelihood strategies on which families depend, from one season to another and from one location to another. This is an important way of spreading risks of livelihood failure in the fragile ecosystem of Lake Chilwa as it enables them to meet their food security requirements, to enhance their incomes and to accumulate wealth, especially among those who are relatively wealthier. However, fishing livelihoods in the plain are vulnerable to the vagaries of weather as much as farming livelihoods are, underlining the strategic importance of natural resource management in the plain and in the Lake Chilwa catchment area.

What is lacking in the LADDER and other studies is a clear typology of how some fishers within the same Lake Chilwa plain advance to greater levels of fishing and wealth accumulation while others are unable to do so. LADDER's findings also cannot be generalised to the Lake Chilwa plain level since the research was conducted only in two villages on the western shore of the lake. Nevertheless, its findings are similar to those obtained from other studies conducted in other parts of the plain and in other African countries (Bene *et al.* 2007, Allison and Mvula 2002). There is also a need to conduct studies that would estimate the multiplier effects of fishing and its contribution to the country's exports and tax revenue so that we can understand its contribution to economic growth at the local and national levels better than we do now. Similarly, at the individual and household levels, there is a need to establish the value-added of full-time fishers arising from fishing by estimating costs, earnings and net profits from full-time fishing and also to estimate the contributions of occasional fishing livelihood strategies to household budgets.

This type of analysis would be essential in understanding the vulnerability of livelihood systems to the shocks experienced by the Lake Chilwa floodplain in the light of climate change and other economic shocks and possible remedies or policy options. Schuijt (1999) estimated that fishing and fish trading on Lake Chilwa generated a value-added of US$17 256 142 per annum in the 1990s. While this estimation is certainly an important methodological contribution to the computation of gross domestic product for the economy of the floodplain, the analysis appears to have excluded the fisheries of the Mpoto Lagoon and the seven rivers that flow into the lake.

8.3.3 Fish trading

Both domestic and cross-border trade in fish have significant multiplier effects in the Lake Chilwa floodplain although their magnitudes are difficult to estimate accurately due to their informal nature. In years of hunger, trading in fish provides supplementary income to food-deficit farmers who resort to fish trading. The poor residents, including women, generally participate in up-stream (post-harvest) fish trading, which has relatively low barriers to entry. They need minimal infrastructure or investment to do this, namely a bicycle, baskets and hooking gear with which to process and transport fish to various markets within the Lake Chilwa basin. Nevertheless, the absence of fresh fish storage or freezing facilities along the shore of the lake results in a high level of fish waste as it is difficult to transport it to profitable markets before it spoils.

Women are involved in fish trading especially in households with adult male fish traders whom they assist on a part-time basis (see Figure 3). Few, if any, women are engaged in long-distance fish trade and their involvement is localised to remote rural set-ups where they combine fish-selling at fish selling points near their homes with domestic chores. In view of the typically high representation of women among the poor, fish trading has the potential to increase their cash incomes. It also enables them to secure some fish for their own consumption, thereby improving the food security and nutritional status of their families. Fish trading can, therefore, spur economic activity in the local economy of the Lake Chilwa plain. However, no studies have been conducted to estimate the extent to which fish trading contributes to the local economy.

8.3.4 Non-fish trading activities

The Lake Chilwa plain has a market system whereby physical markets for both agricultural and manufactured products are organised on a week-day basis. In this system, a particular day of the week is designated as a market day for a particular location or place. The major markets of this nature are at the fish landing points and other developed locations with relatively good road connections, such as Kachulu and Nkuba in the Zomba District, Mposa in the Machinga District and Sombani in the Phalombe District. Other neighbouring markets are found in Govala, Chinamwali, Songani, Domasi and Namwera Turn-Off in the Zomba District and Ntaja, Chikwewo, Nsanama, Chamba and Mpita in the Machinga District. Both agricultural produce and manufactured items produced locally and imported (mostly from China and India) are sold at these markets thus increasing the circulation of cash in the local economy but the extent of this circulation cannot be determined due to lack of studies on this phenomenon. Traders travel all year round from any part of the floodplain to another in pursuit of market opportunities, often trading at centres that operate daily in Mpheta, Domasi, Songani, Namwera Turn-Off, Chamba, Mpita, Ntaja, Jali and Migowi.

Figure 3: Woman processing makumba fish on Lake Chilwa, Source: Kalindekafe (2012)

Prior to the liberalisation of the marketing of agricultural produce in Malawi from 1987, the Agricultural Development and Marketing Corporation (ADMARC) used to operate quite reliable agricultural produce and inputs markets in the Lake Chilwa plain. These markets were held immediately after the harvests of maize, rice and any other crops, and fertilisers are sold in most parts of the plain, especially at markets in Sombani and Migowi in the Phalombe District, Jali, Kachulu, Mpyupyu, Govala, Bimbi and Namasalima in the Zomba District, and Mpheta, Mposa and Ntaja in the Machinga District. However, following the liberalisation, most of these markets were closed down and only a few of them operate today on a rather irregular basis. This has negatively affected the reliability of cash crop marketing and farmers find it logistically difficult and costly to sell their produce to ADMARC, largely due to poor road infrastructure, lack of physical markets and lack of (or limited) means of transport (Kutengule, *et al.* 2006). The situation is particularly difficult during the rainy season, which coincides with the peak hunger season between January and March each year.

The absence of ADMARC markets has created marketing opportunities for vendors who largely buy farm produce cheaply during the harvest season. They buy local produce (especially rice, maize, tobacco and pigeon peas) from urban centres within the country – notably Blantyre, Zomba, Liwonde, Mangochi, Balaka, Lilongwe and Mchinji – and some of it is exported to Zambia and Zimbabwe. The LADDER Project has observed that access to the foreign markets for rice improved the incomes of these traders substantially such that their neighbours became jealous of them (Overseas Development Group 2002b). They only return to sell maize (the staple) to the residents at very high prices during the hunger season, when there is a serious imbalance between high demand and short supply, between December and March. Furthermore, the LADDER Project has documented various problems associated with these private buyers such as dishonesty with their produce weighing scales (*kumanga sikelo*), oligopsonistic pricing of produce (they set the maximum prices for produce as if they were the sellers) and their high selectivity in the crops that they bought. These problems were reported to be factors that limited the profitability of the local produce marketing system at the expense of the farmers thus directly and negatively affecting the viability of farming-based livelihoods. The implication here is that a competitive marketing structure for farm produce would enhance the profitability of farming livelihoods, facilitate an increase in the circulation of money in the local economy, accelerate local economic growth, and raise household cash incomes, thereby contributing to poverty reduction.

8.3.5 Employment opportunities

The diminishing sizes of properties in the plain (due to rapidly increasing population density on land), declining soil fertility, and frequent floods along the major rivers have resulted in low agricultural productivity and production and an increase in households' dependence on casual employment on-farm as well as off-farm. The introduction of irrigation schemes has also increased the demand for casual labour (especially *ganyu*) and therefore casual employment opportunities (Overseas Development Group 2002a and b; Zuka 2009). The fishing industry is another important source of self as well as hired casual employment. The arrival of electricity at most trading centres in the plain has also opened up self-employment business opportunities although no coherent survey of these establishments has been conducted across the plain. However, some people are employed in milling, welding, groceries, bars, mobile phone repairs, video show rooms, and hair dressing salons, which have mushroomed due the availability of electricity. The absence of such a survey is a major gap that has contributed to lack of data and information on the volume of economic activity in the plain.

Owing to lack of large scale commercial enterprises in the plain, there are few formal private sector wage employment opportunities. In this regard, the major sources of wage employment are: government establishments, especially hospitals and health centres; numerous primary and secondary schools; teacher training colleges at Domasi and Liwonde; agricultural extension staff; fisheries officers; community development staff; military, police and prison officers, especially in Zomba; and the National Statistical Office (NSO), the Malawi National Examinations Board (MANEB), and the University of Malawi in Zomba. Non-governmental organisations, such as the Catholic and Anglican churches, also provide employment at their health education facilities within the plain in Zomba City, Malosa, Liwonde, Phalombe, Ntaja and many other smaller places. Although employees do not necessarily come from within the plain, their remuneration trickles down to the local communities and contributes to the circulation of money in the local economies. As with the other economic activities, securing reliable data and information on the size of wage employment in the plain was a challenge. But such data needs to be consolidated at the Lake Chilwa plain level for proper analysis of its impact on the economy of the plain.

8.3.6 Transport sector

Within the plain, transportation is most developed along the main tarmac road from Blantyre to Liwonde, Balaka, Lilongwe and the shores of Lake Malawi, through the city of Zomba. There is significant traffic flow on this road with numerous mini-buses operating between Limbe (Blantyre) and Zomba and between Zomba and the smaller towns of Songani, Malosa and Liwonde – this generates significant value-added that has not been estimated. Similarly, transportation is fairly developed on Lake Chilwa itself with dug-out canoes and engine boats operating between various landing points on the western shore of the lake, the largest of which appears to be the traffic between Kachulu and Chisi Island. There is also some traffic between these landing points and smaller islands and *zimbowela* on the lake and other landing points on the Mozambican side of the lake. Schuijt (1999) estimates the value-added of water transportation on Lake Chilwa at $402 557 per annum, which is a significant contribution to the local economy in terms of the cash incomes earned and the multiplier effects that the open water transportation system has on the wider economy. It also generates employment for boat makers, boat repairers and hired hands, and facilitates the movement of goods and services for purposes of production, distribution and consumption.

The transport infrastructure in the Lake Chilwa plain is poorly developed. The road network is usually in a bad state during the rainy season, preventing the movement of people and goods from one place to another, thereby constraining economic activity and livelihood options available to the people of the area. There is no road for vehicular transport around the western shore of Lake Chilwa. There is not even a tarmac road connection between any of the fish landing points or trading centres along the western shore of the lake and the nearest tarmac road between Blantyre and Lilongwe via Zomba. This poor accessibility by road limits the extent to which the lake can be utilised as an economic resource. Furthermore, although the lake is not far from the major urban areas of Blantyre and Zomba, it offers little or no touristic attraction to Malawians and foreigners due to lack of accessibility and touristic facilities. Even travelling from the western shore of the lake to any of its islands is difficult due to a lack of modern vessels. The poor transport network linking the lake to its catchment area is therefore a major factor constraining the area's contribution to the national economy. In response to poor transport facilities, the relatively new *kabaza* or *dampa* bicycle transport services have been introduced that operate from some points along the tarmac road between Blantyre and Liwonde and major trading centres along the western shore of the lake.

8.3.7 Financial and other services

As with most rural parts of Malawi, there are very few (if any) financial services provided within the Lake Chilwa plain. The most prominent service has recently been introduced in Mwandama village in the Zomba District by the Millennium Villages Project (sponsored by the United Nations Development Programme) whereby a mobile banking outfit has been introduce by Opportunity International Bank of Malawi. Another outlet has just been opened by the NBS Bank at Domasi largely to cater for students of the Domasi College of Education. The LADDER Project found that in some villages, the Malawi Rural Finance Company (MRFC) provided credit services or loans to farmers and off-farm business people although the services were viewed as particularly onerous due to short repayment schedules (Overseas Development Group 2002b). However, since 2005, the Malawi Rural Development Fund (MARDEF) has been providing credit services for farm and non-farm businesses in the plain although access to these loans is limited due to its low coverage. The most reliable banking services for residents of the plain are located in Zomba City, which is quite far away from most parts of the plain. The major banks of the country operate in the city although their services are not accessible to rural poor people principally due to lack of collateral security among the poor.

8.3.8 The structure of the economy of the Lake Chilwa basin

As Table 4 shows, almost all Malawi's economic sectors are active in the plain. However, no figures have been provided, as I could not establish their contribution to the local economy in quantitative terms due to difficulties in assembling data based on the Lake Chilwa plain as a unit of analysis or a coherent economic entity. This highlights the need for surveys to assemble data that can be used to compute the value-added.

Table 4: Lake Chilwa Plain gross domestic product by sector of origin at 2007 prices for the year 2011

Sector	Contribution to GDP in 2011*	Remarks on economic activity or sector and data
Agriculture (small scale and large scale)	?	Gross output data available at Extension Planning Area (EPA)
Fisheries	?	value-added data assembled but focuses on the Lake Chilwa resource
Mining and quarrying	?	Small-scale quarry stone activity near Domasi but no value added data compiled
Manufacturing	?	Active largely as informal sector enterprises but no value added data compiled
Electricity and water	?	Much water produced but no value-added data especially for informal sector water production
Construction	?	Widespread but no value-dded data compiled at plain level
Ownership of dwellings	?	Widespread but no value-added data compiled for both formal and informal sectors
Services	?	Widespread but no value-added data compiled for both formal and informal sectors
Distribution	?	Widespread but no value-added compiled
Transport and communication	?	Widespread but no value-added data compiled at plain level
Financial and professional services	?	Concentrated in Zomba City but no value-added data compiled
Private social and community services	?	Widespread but no value-added data compiled at plain level
Producers of government services	?	Widespread but no value-added data compiled
Unallocable finance charges	?	No value-added data compiled
Total GDP at factor cost	?	No value-added data assembled

* Please note that no complete or reliable data is available for the sectors given above.

The agriculture sector has the best data on gross output compiled at the EPA level and while the data is not value-added it is close to a value-added estimate.

The fisheries sector has been included in the structure due to its prominence in the area. It also has the same degree of data assemblage as agriculture and both sectors have problems of data capture and aggregation of value added. This is especially true for informal or small scale farming and fishing enterprises, the contributions of which to the local economy are difficult to capture in the form of data. The paucity of analysis has important implications for the understanding of changes in the productivity and profitability of the agriculture and fisheries sectors in relation to their potential to contribute to improvements in household incomes, increases in the local purchasing power, and the wider multiplier effects that could boost the local and national economies. This implication applies to all the other sectors.

8.4 CONCLUSION

The analysis in this chapter has showed that there is an intricate relationship between the environmental and economic processes in the context of climate change, which contributes to fluctuations in the level of Lake Chilwa, with some ramifications for the economy and livelihoods of the people of the plain. There is a need to analyse both processes coherently so as to generate insights into how the extraction of natural resources for human survival and economic prosperity can be sustained over a long-term horizon and inter-generationally. Lack of reliable data and information on the two processes inhibits this type of analysis.

In relation to the economic structure I described above, the natural resource base of the Lake Chilwa plain provides the residents of the plain with opportunities to undertake various economic activities and livelihood strategies. The prominent ones are in rice farming, fishing, wage and casual employment, and trading in fish and non-fish items. There is significant potential for these economic activities and the tourism industry to contribute to the growth of the local and national economies should there be any increase in physical infrastructural development (especially of roads and irrigation), natural resource management, and the delivery of vital economic and social services such as banking, electricity, health, education and water.

The absence of data and information on the size of the economy of the plain makes it difficult for one to appreciate the contribution of various livelihood activities or economic activities to the local and national economies. For a robust analysis of the local economy to be undertaken, one needs some data sets on each of the sectors of the local economy to allow one to compare them to national sector figures and compare constraints to growth between local and national economies. While the structure of the local economy is comparable to that of the national economy its size is difficult to estimate unless the relevant data is either collected or assembled. The structure points to potential sources of growth or livelihood that government can utilise optimally with relevant developmental policies, economic and social infrastructure, responsive and responsible institutions, and improvements in the functioning of markets.

In this chapter, I described the intricate relationship between fishing and non-fishing activities and how it creates multiplier effects for the economy of the area. There is also a complex relationship between the natural resource base, farming, fishing and other livelihood activities in the Lake Chilwa plain. Poor management of the common property resources is creating pressure on the livelihoods of the people of the area and this makes it difficult for poverty reduction to materialise for the locals as well as recent migrants. The in-migration of the rural poor into the seemingly open-access/unrestricted fishery of the Lake Chilwa plain appears to have brought about the 'tragedy of the commons' with serious over-fishing.

In the final analysis, the complex links between fishing and other livelihood activities call into question the categorisation of fishing into either a 'last-resort' or 'first-resort' livelihood strategy in response to the desire to avoid poverty or to accumulate wealth. Livelihood activities ought to be viewed as a continuum of attempts to address both survival and wealth accumulation strategies that are essentially inseparable owing to the fragile ecosystem of the Lake Chilwa plain. This justifies the need to address the challenges of fishing livelihood strategies in a multi-sector context, which has not been the case in Malawi. The recent integration of the Fisheries Department into the Ministry of Agriculture, however, appears to be a positive development towards realising those links.

REFERENCES

Acharya SN. 1978. 'Perspectives and problems of development in low income sub-Saharan Africa.' *World Bank Staff Working Paper No. 300.* World Bank, Washington DC.

Allison E and Mvula P. 2002. 'Fishing livelihoods and fisheries management in Malawi.' *LADDER Working Paper No. 22.* University of East Anglia, United Kingdom.

Barth R and Hemphill W. 2000. *Financial Programming and Policy: The Case of Turkey.* International Monetary Fund, Washington DC.

Bene C Macfadyen G and Allison E. 2007. 'Increasing the contribution of small-scale fisheries to poverty alleviation and food security.' *Fisheries Technical Paper No. 481.* Food and Agricultural Organisation of the United Nations, Rome.

Chiotha S. 1996. 'Lake Chilwa dries up: Need for management plan.' *Paper presented at a Workshop on Lake Chilwa Management Plan organised by the Malawi Government (Fisheries Department) and the University of Malawi.*

Cornia GA, Jolly R and Stewart F (Eds). 1987, *Adjustment with a Human Face Volume 1: Protecting the Vulnerable and Promoting Growth.* Clarendon Press, Oxford.

Cornia GA and Helleiner GK (Eds). 1994. *From Adjustment to Development in Africa: Conflict, Controversy, Convergence, Consensus?* MacMillan Press Ltd, Basingstoke and London.

Government of Malawi and United Nations. 1993. *Situation Analysis of Poverty in Malawi.* UNICEF, Lilongwe, Malawi.

Gupta D. 1974. 'Industrial development in Malawi', *Journal of Social Science* 3: 35-47.

Humphrey DH. 1973. 'Malawi's economic progress and prospects', *Eastern Africa Economic Review* 5(2): 71-104.

Kalindekafe MP. 2010. 'The Chilwa Basin ecological and environmental status in relation to pollution, siltation, biodiversity and habitat change.' *Draft paper prepared for the DARMA Research Programme, July 2010.*

Kalk M, McLachlan A and Howard-Williams M (Eds). 1979. 'Lake Chilwa: Studies of change in a tropical ecosystem', *Monographie Biologicae* 35: 17-227.

Kutengule M. 2000. 'Farm and non-farm sources of income: Rural livelihood diversification in Malawi.' *Unpublished PhD thesis.* University of East Anglia. Norwich, England.

Kutengule M, Nucifora A and Zaman H. 2006. 'Malawi: Agricultural Development and Marketing Corporation reform' pp415-451 in: Coudouel A, Dani A, and Paternostro S (Eds) *Poverty and Social Impact Analysis of Reforms Lessons and Examples from Implementation.* World Bank, Washington DC.

Kydd J and Christiansen R. 1982. 'Structural change in Malawi since independence: Consequences of a development strategy based on large-scale agriculture', *World Development* 10(5): 355-375.

Lancaster N. 1979. 'The changes in the lake level', pp. 17 – 227, in: Kalk M, McLachlan A and Howard-Williams C (eds) *Lake Chilwa: Studies of change in a tropical ecosystem. Monographie Biologicae,* Vol. 35,. Dr W. Junk Publishers, The Netherlands.

Malawi Government. 1994. *Survey of Household Expenditure and Small-scale Economic Activities 1990/91.* National Statistical Office, Zomba.

Malawi Government. 2006. *Annual Economic Report 2006.* Ministry of Economic Planning and Development, Lilongwe, Malawi.

Malawi Government. 2011. *Annual Economic Report 2011, Budget Document No.2.* Ministry of Development Planning and Co-operation, Lilongwe, Malawi.

Malawi Government. 2000. *Annual Frame Survey Report.* Fisheries Department, Lilongwe.

Malawi Government. Undated. *National Sample Survey of Agriculture 1992/93 Volume II, Smallholder Garden Survey Report.* Zomba, Malawi.

National Statistical Office (NSO). 1998. *Malawi Population and Housing Census 1998, Preliminary Report.* NSO, Zomba, Malawi.

National Statistical Office (NSO). 2008. *Malawi Population and Housing Census 2008, Preliminary Report.* Zomba, Malawi.

Njaya FJ. 2002. 'Review of management measures for Lake Chilwa.' *Project paper submitted to the United Nations University.* UNU, Reykjavik Iceland.

Overseas Development Group (ODG). 2002a. *LADDER Village Report No. 26: Katanda Village, Malawi.* University of East Anglia, Norwich. Available at: *www.odg.uea.ac.uk/ladder_(23)* (June 2012).

Overseas Development Group (ODG). 2002b. *LADDER Village Report No. 27: Sauka Village, Malawi.* University of East Anglia, Norwich. Available at: *www.odg.uea.ac.uk/ladder* (October 2011)

Peters P. 2004. *Informal Irrigation in Lake Chilwa Basin: Stream-bank and Wetland Gardens. Broadening Access and Strengthening Input Market Systems.* Collaborative Research Support Programme (BASIS-CRSP), University of Wisconsin, Madison. Available at: *http://www.basis.wisc.edu* (April 2012).

Pryor FL. 1990. *The Political Economy of Poverty, Equity and Growth: Malawi and Madagascar. World Bank Comparative Study.* Oxford University Press, Oxford.

Turpie J, Smith B, Emerton I and Barnes B. 1999. *Economic Value of the Zambezi Basin Wetlands.* IUCN Regional Office, Cape Town.

Salama, R and Jones D. 1982. 'Fish exports from Malawi – a supply study.' *Malawi Export Promotion Council Desk Research Study Report.* Malawi Export Promotion Council, Blantyre, Malawi.

Schuijt KD. 1999. 'Economic valuation of Lake Chilwa Wetland.' *State of the Environment Study No. 18.* Lake Chilwa Wetland Project, Zomba, Malawi.

Vauhan M. 1982. 'Food production and family labour in southern Malawi: The Shire Highlands and Upper Shire Valley in the early colonial period', *Journal of African History* 23(3): 351-364.

White L. 1987. *Magomero: Portrait of an African Village.* Cambridge University Press, Cambridge.

World Bank. 1989. *Sub-Saharan Africa: From Crisis to Sustainable Growth: A Long-term Perspective Study.* World Bank, Washington DC.

World Bank. 1990a. *Making Adjustment Work for the Poor: A Framework for Policy Reform in Africa.* World Bank, Washington DC.

World Bank. 1990b. *Malawi: Growth through Poverty Reduction.* Report No140-MAI. World Bank, Washington DC.

World Food Programme (WFP), USAID/FEWS Project and Malawi Government. 1996. *A Quest for Causality, Vulnerability Assessment and Mapping (VAM): Malawi baseline 1996.* Lilongwe, Malawi.

Zuka SP. 2009. 'Theory, practice and outcomes of community management of common property resources through the lenses of Njala and Domasi irrigation schemes.' *Unpublished Masters thesis.* University of Malawi, Zomba, Malawi.

Chapter 9

TOWARDS DEFRAGMENTING MANAGEMENT OF LAKE CHILWA BASIN RESOURCES: ISSUES AND CONFLICTS

Friday NJAYA, PETER MVULA KALINDEKAFE

9.1 INTRODUCTION

In this chapter, we will analyse the state of the environment of the basin by examining integration of ecological, social and economic information based on the Driver–Pressure–State–Impact–Response (DPSIR) framework, which is a general framework for organising information about state of the environment as outlined in Chapter 1. We also assess the degree of fragmentation and explore ways of defragmenting the management system. As the preceding chapters have indicated, several sectors exist in the Lake Chilwa basin especially from the colonial times, when fragmentation of the resource management first occurred. The key resources commonly found within the basin that are of economic benefit to the local communities include fish, birds, forest, land, grass (grazing) and water. Several sectors or sub-sectors exist in the basin based on particular common pool resource that are of particular interest to them. However, we will first examine the status of the environment before looking on the interactions between the resources and the various actors.

9.2 CHANGES IN THE STATE OF THE ENVIRONMENT OF THE BASIN

Based on the DPSIR framework (Carr *et al.* 2007) and issues highlighted in the preceding chapters, increase in human population has an enormous impact on environmental degradation in the Lake Chilwa basin. In this analysis, *driving forces* include socio-economic and socio-cultural forces driving human activities, which increase or mitigate pressures on the environment, such as human population changes within the basin. The *pressures* relate to stresses that human activities place on the environment, such as pollution from using chemical fertilisers. The *state of the environment* refers to the condition of the environment, for which we may use changes in biodiversity as an indicator. The *impacts* are the effects of environmental degradation, notably food availability and increased vulnerability to diseases. Finally, the *responses* refer to formal policies and informal coping mechanisms used by the resource user communities to take care of the environmental situation.

In his economic analysis, Shuijt (1999) noted that the Lake Chilwa wetland is dominated by: fish (*Barbus* species, *Clarias* species and *Oreochromis shiranus*); forestry products for use as fuel wood, building materials, fuel for fires for fish processing, and timber for making canoes and boats; vegetation products for making brooms, mats, fish traps, baskets and building materials; water for irrigation, transport and domestic use; and land to grow cash crops like cotton, rice and tobacco, and maize and vegetables for subsistence purposes.

9.2.1 Demographic changes

In this book, we assessed pressures by looking at demographic changes in the basin. The Lake Chilwa catchment area has approximately 1.6 million people with a density of 321 people per km^2 as compared to 104 people per km^2at national level, representing an increase of approximately 50% from that of 1998 (NSO, 2008). These figures indicate that the Lake Chilwa catchment is one of the most densely populated areas in Malawi. As highlighted by Kalindekafe and Mvula in Chapters 2 and 3 respectively, the local community of the area has largely depended on natural resources for survival since they settled in the basin, especially from the 17th century onwards.

Around 1.6 million people are involved in both rain-fed and irrigated farming systems within the basin. They grow crops like maize, rice, groundnuts, sorghum, tobacco, millet, cassava and potatoes for both cash and food. Intensive farming along river banks and in marginal areas is common, to the extent that soil erosion and siltation are threatening aquatic life including fish in rivers and the lake. Government set up irrigation schemes in the 1960s and 1970s, which were declared public land, and fertiliser run-off from these schemes further has contributed to the increase in chemical pollutants in the lake, as well as contributed to population growth.

Nearly 6 000 fishers are engaged in fishing on the lake (open waters), rivers and lagoons of the Lake Chilwa basin, contributing between 25% and 30% of the total fish caught in Malawi. It is estimated that nearly 1.2 million birds are trapped every year by 461 bird trappers (Maloya 2001). In terms of wildlife, birds mainly water fowl are an important source of protein as well as income. At the beginning of the 20th century, wildlife was commonly found within the Lake Chilwa basin, such as hippos, kudus, sable, antelope, bushbuck, reedbuck, hartebeest, lions, hyenas and leopards. However, these animals have become scarce, with some, like lions, becoming altogether extinct due to human population pressure and activity (Howard-William 1973). Forestry resources are also under pressure, as many people venture into uncontrolled charcoal production. Lake Chilwa and its wetlands are not protected so people have open access to the resources of the area. The local community regards wood as a *free good* and there is almost no alternative form of fuel for domestic cooking or smoking fish (Kalk 1979). The decline in the resources is generally attributed to change of the environment and an increase in human population with accompanying destructive utilisation of the natural resources.

Mvula, Njaya and Kambewa in Chapters 3, 4 and 5, respectively, argue that due to the population growth and increase in demand for access to land and fishing rights, there has been an 'elite capture' phenomenon. This entails benefits that the traditional leaders obtain by earning an income through tributes. This has been happening, as immigrant land seekers and fishers require paying the village heads a specified amount of money for the rights to access land or fish from the lake. This is possible because the resources requiring tribute in fact are excludable. This implies that better and more just rules for access are possible.

Kambewa in Chapter 5 outlines conflicts over access to *dambo* (farm land) that emerge due to increase in population. He states that there have been conflicts over access to *dambo* areas mainly due to population pressure, as there is a serious land shortage problem in the area. In addition, even where people have land, the rainfall is often erratic which results in low yields in both low lying and upland areas of the catchment area. *Dambos,* therefore, remain good areas for producing food crops such as maize, peas and potatoes. However, he notes, that *dambo* land is relatively inadequate to support the population and its various productive activities in the area. Consequently, some people have resorted to cultivating their crops on stream banks which results in high soil erosion and subsequent siltation of rivers and the lake. He further states that there have been conflicts between farmers and environmental experts over the cultivation of land along stream banks. The agricultural and environmental experts recommend that there should be no cultivation at least eight metres from the river, but most farmers usually ignore this rule.

Kutengule in Chapter 8 presents attributes of the increase in population in the Lake Chilwa basin. The catchment area on the Malawi side is 5 669 square kilometres while that on the Mozambique side is 2 680 km^2. The high population increase is attributed to high fertility rate and migration into the area from other parts of the country and Mozambique. Njaya (2002) reports that fishing activity on Lake Chilwa has increased tremendously between the mid-1970s and the late 1990s as evidenced by increases in fishing gear, craft, number of gear owners, crew members, boats, traps, and immigrant fishers from Lakes Malombe and Chiuta.

Similarly, the LADDER Project documents population movements from one part of the western shore of the lake to another, principally in search of farming and fishing opportunities. These movements were mostly from the high population density district of Phalombe to the Zomba part of the western shore as well as from Mozambique into Malawi (Overseas Development Group 2002b). The increase in the fishing effort, a rise in population movements, increasing population density on farm land, and the closure of employment opportunities on South African mines, have combined to exert pressure on the fishery of Lake Chilwa with negative effects on the livelihoods of the people of the area. In effect, there is diminishing agricultural productivity and production due to increased pressure on the land; diminishing fish catch and earnings from fishing; conflicts between indigenous residents and immigrants over fishing rights, farm land, and grazing land; an increase in bribery and corruption over access to fisheries, land and other natural resources; and a rise in the incidence of infectious diseases, especially HIV and AIDS.

9.2.2 Pollution and increase in use of chemical fertilisers

To examine pressures, pollution and use of chemicals were used as indicators. Major sources of pollutants into the Lake Chilwa include agriculture (rain-fed and irrigation) that use pesticides and fertilisers, deforestation that encourages topsoil and organic matter erosion and Zomba city that releases organic and inorganic pollutants to the Lake Chilwa basin as presented in Chapter 2 by Kalindekafe. With the increasing population pressure and agricultural activities in the catchment, pollution levels are likely to increase. The influent Likangala and Domasi rivers showed that river flow, conductivity and levels of total suspended solids (TSS) in the water affect upstream migration of *Barbus sp (matemba* fish). In addition, increasing deforestation and conversion of agricultural land to homestead development leads to erosion and sedimentation, particularly in the Likangala catchment basin (Jamu *et al.* 2006). Lake Chilwa receives water from Domasi, Likangala, Namadzi, Sombani, Thondwe, and Phalombe Rivers from the Malawi side where the water is being used for different purposes such agricultural production as has been shown above. Another example, the increasing use of water for industrial and other productive activities in the upstream area of Likangala River creates numerous problems for downstream communities. The establishment of an abattoir near the Likangala Bridge, and the leakages from the sewage disposal systems by Zomba Central Prison, Zomba Central Hospital, Cobbe Barracks and Zomba Mental Hospital has resulted in the dumping of waste into the river. These activities contribute to the pollution of the water of the river and consequently, the wetland (Mulwafu 2000).

9.2.3 Environmental status

Assessment of the status of the environment mainly relates to change in geology of Lake Chilwa basin, biodiversity (fish, birds and land holding sizes) and to a certain extent, water quality in rivers and lakes. Kalk *et al.* (1999) observe that the geological changes, which happened before the formation of the sandbar between Lakes Chilwa and Chiuta about 9 000 BC, led to the present form of the Lake Chilwa basin as one of the closed drainage lakes. In terms of the fishery, Kutengule in Chapter 8 highlights the status of fishery resources in Lake Chilwa as it also affects national production levels. During recession times national fish production goes down; for example, in 1995 when Lake Chilwa dried up, its annual catch was only 1 328 metric tons, which was only 4% of total national fish production. This decline in national fish production, as he observes, is a result of the Lake Chilwa recession, which underlines the strategic importance of the lake in national fish supply and economic and nutritional well-being.

Kutengule in Chapter 8 also observes that there have been declining employment opportunities due to diminishing landholding sizes in the plain (due to rapidly increasing population on land), declining soil fertility, and frequent floods along the major rivers which have resulted in low agricultural productivity and production and an increase in households' dependence on casual employment on-farm as well as off-farm. On the other hand, the introduction of irrigation schemes has also increased the demand for casual labour (especially *ganyu*) and therefore casual employment opportunities.

Kalindekafe in Chapter 2 examines bird population changes. Many species of birds breed during the months January to July at various sites around the lake (Wilson 1999). Most birds favour the river mouths for breeding, as these areas are largely inaccessible to the local people and offer protection to the breeding birds because of the dense marsh vegetation, (Bhima, 2006). Bird species that are trapped in large numbers include common moorhen *Gallinulachloropus*, lesser moorhen, Allen's gallinule, Blake crake, fulvous whistling-duck, white-faced whistling-duck *Dendrocygna viduata* and Hottentot teal *Anas hottentota* (Wilson and van Zegeren 1998, Wilson 1999). There are 461 bird trappers operating along Lake Chilwa beaches (Environmental Affairs Department 2000). When the water level drops, catches of birds greatly increases by about 300-500% (Bhima 2006).

9.2.4 Food and diseases

In identifying impacts, indicators used include declining food availability and increased vulnerability to diseases as outlined in Chapter 8 by Kutengule. Annual crop yields have generally been declining from the 1990s due to erratic rainfall which resulted in flooding and droughts; pest attacks; declining soil fertility due to lack of crop rotation in the face of rapid population growth; labour constraints; rising costs of farm inputs; inefficient input and produce marketing largely due to poor road infrastructure and lack of agricultural extension and credit services. The implication here is that addressing these constraints would foster agricultural production in the plain, thereby boosting its contribution to the national economy, food security, and socio-economic well-being at both local and national levels. From the year 2004, however, maize production and food security have improved in the plain due to a combination of good rainfall and greater availability of subsidised fertilisers and seed. The other constraints need some consideration in order to raise agricultural productivity and production.

In addition to crop production, livestock, especially chickens, ducks, goats and cattle are kept by the people of the plain (Kutengule 2000, Overseas Development Group 2002a and b). Various payments or user charges were introduced on services that were previously free of charge. People could not afford these charges and the result was that their livestock could not get proper services. This reduced the survival rate of livestock as it become vulnerable to disease eventually succumbing to diseases. Livestock production was further affected by the lack of veterinary assistants in the rural areas, the closure and partial privatisation of dip tanks and veterinary services, cattle rustling (thefts) and diminishing grazing land.

There has been a rise in the incidence of various diseases such as cholera and infectious diseases, especially HIV and AIDS within the Lake Chilwa basin. Other diseases include bilharzia and malaria. Cholera usually breaks out due to use of untreated water and lack of health facilities around the lake. HIV and AIDS incidences are common due to common use of transactional sex among and between male fishers and female fish traders in fishing camps concentrated around the lake.

9.2.5 Policies and rules

Responses were assessed by looking at existence of formal policies and informal rules that Chapters 4 and 7 by Njaya and Chinsinga, respectively outline. In addition, on-going and past programmes and projects were also identified by Kalindekafe in chapter 2. Table 5 below shows various policies and strategies that are applied in the management of Lake Chilwa resources.

Table 5: Key national sectoral policies and strategies applied in the management of the resources in the Lake Chilwa basin

Policies/Strategies	Objectives/Focus areas
Malawi Growth and Development Strategy II (MGDS)	The main thrust of the MGDS II of 2012 is to create wealth through sustainable economic growth and infrastructure development as a means of achieving poverty reduction. Within the MGDS II, the medium term outcome is 'to ensure sustained fish availability for food and nutrition security as well as income generation for small and large scale fisheries'.
National Environmental Policy	The National Environmental Policy of 2004 calls for development of mechanisms for cross sector management; facilitating active participation of local communities and other stakeholders in enforcement of legislation; and integration of environmental planning, management and institutional frameworks into the decentralised structure.
National Fisheries and Aquaculture Policy	The goal of the National Fisheries and Aquaculture Policy of 2001 aims at achieving optimal exploitation and utilisation of the fisheries of Malawi's water bodies and promotion of investment in both capture and culture fisheries.

Irrigation Policy	The Irrigation Policy of 1996 aims at promoting social and economic development through irrigated agriculture that is sustainable over time, economically justified, financially viable, socially acceptable and technically spent, without causing unacceptable impacts on the environment.
National Land Resource Management Policy	The National Land Resource Management Policy of 2000 aims to promote the efficient and diversified and sustainable use of land resources both for agriculture and other uses in order to avoid sectoral land use conflicts and ensure sustainable socio-economic development.
National Forestry Policy	The National Forestry Policy of 1996 focuses on sustaining the contribution of the national forest resources to the quality of life in the country by conserving the resources for the benefit of the nation.
Wildlife Policy	The Wildlife Policy of 2000 aims to ensure proper conservation and management of wildlife in order to provide for sustainable utilisation; equitable access to the resources; and fair sharing of the benefits from the resources for both present and future Malawians.
Water Resources Policy	The overall goal of the National Water Resources Policy (2005) goal is sustainable management and utilisation of water resources, in order to provide water of acceptable quality and of sufficient quantities, and ensure availability of efficient and effective water and sanitation services that satisfy the basic requirements of every Malawian and for the enhancement of the country's natural ecosystems.
Malawi Decentralisation Policy	The Malawi Decentralisation Policy of 1998 seeks to create a democratic environment and institutions in Malawi for governance and development at the local level which will facilitate the participation of the grassroots in decision making; eliminate dual administrations (field administration and local government) at the district level with the aim of making public service more efficient, more economical and cost-effective; promote accountability and good governance at the local level in order to help Government reduce poverty; and mobilise the masses for socio-economic development at the local level.
Gender Policy	The Gender Policy of 2008 seeks to mainstream gender in the national development process in order to enhance participation of women and men, girls and boys for sustainable and equitable development.
Malawi National HIV/AIDS Policy	The goal of the Malawi National HIV/AIDS Policy of 2003 is to prevent HIV infections, to reduce vulnerability to HIV, to improve the provision of treatment, care and support for people living with HIV and AIDS and to mitigate the socio-economic impact of HIV and AIDS on individuals, families, communities and the nation.

9.3 Rules and regulations

Traditional authorities apply various rules in the management of the resources especially fish, water forest and land in the Lake Chilwa basin within co-management arrangements as outlined in Chapters 3 to 7 by Mvula, Njaya, Kambewa, Mulwafu and Chinsinga, respectively.

Under the community-based natural resource management, village heads and chiefs preside over specific areas. Village heads and chiefs can grant access rights to resource users within their specific jurisdiction. In this regard, the village heads become important access points to livelihood resources such as land, fish and wildlife (birds). Their control over access to these resources enhances the social status of chiefs over ordinary villagers. At the time of Malawi's independence in 1964, the new government maintained the powers of the chiefs and village heads over land and other resources, subject to the overall authority of sectoral departments. For example, the establishment of irrigated rice settlement schemes in the floodplain and the increasing opportunities that fishing and fish trading offered in the basin were only possible because sectoral departments were involved. The explanation of how chiefs appropriated common land as private property lies in their ability to use their traditional powers to accumulate land to themselves hence they could distribute to either residents or new migrants.

In terms of fisheries, chiefs in lakeshore districts were empowered to regulate access to lakes. The traditional leaders could limit access to the fishing grounds in the fisheries-related activities in form of levies as asserted by Chirwa (1996). Each community claimed access to the beaches closest to them, and membership to the community was quite an important determinant of access. There are several local institutions that govern access to exploitation and preservation of the fisheries. These are monitored and sanctioned by the local leaders. The chiefs could levy taxes on canoes and quotas imposed on trees to be cut for canoes. It is however unclear whether the revenue collected was used for managing fisheries resources.

9.4 SCALE OF THE FRAGMENTATION

In this section we will look at the various resources and the entities that have interests in each of the resources and consequently the conflicts that these interests bring about. It is clear from Chapter 1 that there is a lot of interest and activity in the following resources: fish, land, water, birds, fuel wood, pasture and forest. These resources also suffer greater fragmentation in terms of management. It is clear from that chapter that of these five, fish, land, and water generate a lot more interest from many parties including resource users, government departments, non-governmental organisations and the academia. Table 6 shows the resource versus policy interactions.

Degree of influence is assessed by looking at the number of *X*s. Many *X*s means the stronger the influence of that organisation in that resource. It is clear that there is a lot of interest in fish from the Department of Fisheries, International Organisations, the academia and some influence from health sector, the Department of Parks and Wildlife, and the Department of Energy. All these have their own interest and employ different strategies which in the long run bring about conflicts in terms of how management of this sector is to be done. Regarding land,

Table 6: Degree of influence of the relevant sectoral policies on specific natural resources in Lake Chilwa basin

Natural Resource	Policies influencing utilisation of the key resources										
	Fisheries Policy	Forestry Policy	Agriculture Policy	Wildlife Policy	Irrigation Policy	Decentralisation Policy	Land Resource management Policy	Environmental Policy	HIV/AIDS Policy	Gender Policy	Water Resources Policy
Fish	XXX		XX	X	X	X	X	XX	XX	XX	XXX
Birds		XX		XXX		X		XX	XX	X	X
Water					XXX	X	XXX	XX	XX	XXX	XXX
Grass				X	X	X	X	XX	XX	XXX	X
Trees		XXX	XX	XX		X	X	XXX	XX	XXX	X
Land			XXX		XXX	X	XXX	XX	XX	XXX	XXX

Note: X = weakest influence XX= moderate influence XXX = strongest influence

the sectors with a strong influence are agriculture and irrigation, but as chapters 4 and 6 by Njaya and Mulwafu, respectively, show, water and fisheries sectors have an interest too because the agricultural activities impact negatively on activities of their sectors, i.e. the pollution of the water, the abstraction of the water and the poisoning of the fish bring the two former sectors in conflict with the latter two.

Water too draws interest from the department of irrigation, agriculture, international organisations, academia and the department of health. As Mulwafu shows in his chapter, in the quest to achieve food security the departments of agriculture and irrigation are advocating for the use of fertilisers and irrigated farming. These practices end up poising the fish in the lake, thereby suffocating the fishing sector because there is less fish available. Furthermore, through damming and channelling, agriculture is taking up the water that is needed by the fish downstream. In addition, there is a lot of interaction between birds and the sectors of parks and wildlife. The international conventions like RAMSAR and Convention on Biodiversity also influence conservation of some resources such as fish, water and birds as outlined in chapter 2 by Kalindkafe. This poses a dilemma for management and governance which may lead to potential conflicts between the resource users and natural resource conservation experts.

9.4.1 Gender

Women in the Lake Chilwa wetland, just like in the country as a whole have a higher representation than men. Binauli and Chipeta (1999) established that division of labour that exists in the wetlands emphasises women's role in natural resource utilisation and management. Women generally spend much of their time collecting fuel wood, water, and working on the land. The two authors further established that traditional roles of men and women are in place, especially with regard to homes where it is the women's job to collect water and firewood and raise children. It is the men's role to carry out fishing and house building. These are clear gender roles, but some work such as tending the maize garden and fish processing is shared. While women take an active role in food production, it is normally the men who will sell any surplus produce and take the proceeds. It was also established that even though the matrilineal system dominates in the area as Mvula in Chapter 2 has also demonstrated, it is the man who is the source of authority. Women tend to be domineered and have a subservient role, maintaining the home and catering for the husband and children. The same applies to decision making when it comes to what to grow on the land which effectively falls under the ownership of men.

9.4.2 HIV and AIDS

A further dimension of poverty and vulnerability in fishing communities lies in the common perception of fisherfolk's dissolute lifestyles. Allison and Mvula (2002) indicate that tales of habitual drunkenness and expenditure of substantial portions of cash income by fishermen on beer and prostitutes (and consequent high risk of being infected with and spreading HIV). They indicate that women in fishing communities are thought to be particularly vulnerable to discrimination and neglect. Allison and Mvula (ibid) in their work around the Lake Chilwa established that there were links between fishing and other rural services and that fishing had stimulated in the shores of Lake Chilwa a number of economic ventures. They also established though that profits from fishing were also directed to unproductive spending: 'An average fisherman [on Lake Chilwa] earns £150 per annum and most of this money is spent locally, especially on beer and women' (Chilivumbo in Kalk 1979: 49). Similarly, in 2001, distilling, selling and consuming *kachasu*, which contains health-threateningly high levels of methyl alcohol, is reported as a significant livelihood activity in most fishing villages and promiscuous migrant fishermen are implicated in the rapid spread of HIV and AIDS in the lakeshore area.

9.5 KEY ISSUES AND CONFLICTS ASSOCIATED WITH THE USE AND MANAGEMENT OF THE RESOURCES

The various chapters in this book have pointed to the fact that due to the fertility of the land around the floodplain and the fish that the lake provides, there has been an increase in the population of the area (jumping from 162 persons per km^2 in the 1980s to 321 persons per km^2 in 2008) which has led to intensive farming including farming in marginal areas: swampy areas, birds sanctuaries, river banks and hills. This population boom, coupled with the presence of so many sectors and organisations that have interest in the management of the various resources, is a source of conflicts that manifest themselves in various forms. Some key conflicts that have been highlighted in the papers in this volume include resources access, governance, policies, ecological issues, transboundary and agricultural development.

9.5.1 Governance issues

Increase in human population in and around the areas of the wetland has not spared the wetlands with regard to conflicts over water because demand for water has gone up. The increase in population is mainly shown within the fisheries, agriculture and water sectors. Fishers from other lakes especially Lake Malombe migrate to Lake Chilwa either on seasonal or periodic basis in search of good landings. Reviews in this volume also show how populations downstream of rivers that feed the wetland are increasingly in conflict with those in the upland because the population in the upland areas now use more water and thus less water goes downstream. Additionally, conflicts in domestic water use have extended to the level of households whereby people from one household would claim ownership to a source and bar others from accessing it or simply ignore laid down procedures insisting to access water before those that queued first.

Development in a form of tobacco estates and rice irrigation schemes have meant the damming of rivers or construction of canals, that draw water from rivers that feed the lake and thus depriving communities downstream. Much as the two activities bring in the much needed foreign exchange and contribute to national food security, the fact of the matter is that people who have always depended on these rivers for their livelihoods suddenly find themselves in a situation where water has become a scarce commodity.

Weak governance of the resources due to unclear roles of various actors, limited empowerment of the resources users, unrecognised reforms within the decentralised natural resource management and limited participation, accountability and transparency has also led to conflicts. Conflicts in fisheries management have arisen as a result of the establishment of legal entities in the form of Beach Village Committees which are fisheries regulatory bodies at the community level. These are in conflict with the traditional leadership which has for a long time, assumed the role of managing the fishery by having the authority to allow immigrants to fish in the lake. The traditional leadership get a token of thanks from the people that they allow onto the lake. The BVCs get on a collision course with them when they question the practice.

9.5.2 Institutional issues

There is also the conflict between policies in that various entities with interest in the different commons in this area use different strategies which sometimes rub against each other. A good example is that of the ministries of irrigation and agriculture who in advancing the food security agenda encourage people to cultivate on the river bank. This type of cultivation has the potential of causing siltation of the lake due to soil erosion and this consequently has negative impact on spawning grounds of fish. Thus, apart from being in conflict with fisheries sector, the policy violets the National Environmental Policy of 1996.

The scientific network so far established in the DARMA project should be encouraged and a mechanism of sustainability formulated. This will ensure a holistic approach to development of management plans and strategies for the improved livelihood systems of the basin communities.

9.5.3 Ecological issues

Seining in sanctuary areas where birds breed is highly contested between fishers and local residents that hunt for income and food. In some cases the sanctuary areas are referred to as 'sacred' places where, according to customary arrangements, no disturbance on the ecosystem is allowed. While the resident fishers observe the sanctuary areas, the in-migrant *nkacha* seine fishers from Lake Malombe do not. This is also a potential area of conflict that exists between the bird hunters and the in-migrant fishers. Pollution due to agricultural activities and city wastes has greatly deprived communities downstream of clean water, leaving them bitter and helpless at the same is also another potential area of conflict.

9.5.4 Transboundary conflicts

The Lake Chilwa wetland and catchment area is shared between Malawi and Mozambique such that the wetland fishery is under threat from overfishing and siltation of breeding sites in the main influent rivers from the two countries (Jamu *et al.* 2006). The major influent rivers include Domasi, Likangala, Phalombe and Sombani from Malawi and Mnembo and Chimazi from Mozambique. Fishing-related conflicts regarding the shared ecosystem as outlined by Njaya (2001) arise due to different management systems with Malawi adopting co-management regime and on the Mozambican side having traditional based management system. Subsequently, fishing regulations such as closed season, mesh size regulations and fishing practices that are observed on the Malawian side are non-existent on the other side. Apart from fish, water is also another contested resource as noted by Ohlsson (1995). This potential area of conflict demands adoption of a transboundary approach in the management of the resources.

9.5.5 Technological issues

There are also conflicts over methods of exploiting various resources as chapter 4 by Njaya shows. In fishing some operators especially the in-migrant fishers have introduced *Usodzi wa Mululu* whereby weeds are cleared within a fishing ground before seines are used. In this case there is a threat on some biological processes mainly in relation to breeding and growth of *Oreochromis* species as indicated by Kalk *et al.* (1979). The fishing method is highly challenged by the local fishers, and hence, the conflicts.

9.6 CONCLUSION

The state of the environment and challenges and conflicts are a manifestation of the importance of the resources found in the Lake Chilwa basin, and the lack of a common or concerted effort in managing them. This calls for an ecosystem-based approach with risk assessment to the management of the resources. In this case decentralised management of the resources in an ecosystem-based approach is recommended. This will ensure that the Lake Chilwa basin CPR management systems is defragmented with participation of the resource users and other key stakeholders including the government departments, academic/research centres and non-state actors. Of particular importance are the issues on social development and decent work that recognise employment safety, gender and child labour. Climate change is another dimension that needs more research as it impacts on the resources within the basin.

REFERENCES

Allison E and Mvula P. 2002. 'Fishing livelihoods and fisheries management in Malawi.' *LADDER Working Paper No. 22*. University of East Anglia, United Kingdom.

Bhima R. 2006. 'Subsistence use of water birds at Lake Chilwa, Malawi', pp255-256, in: Boere GC, Galbraith CA and Stroud DA (eds) *Water Birds around the World*. The Stationery Office, Edinburgh, United Kingdom.

Binauli L. and Chipeta L. 1999. *State of the Environment Study No. 1, Lake Chilwa Wetland and Catchment Management Project*. Government of Malawi /DANIDA, Zomba, Malawi.

Carr E, Wingard PM, Yorty SC, Thompson MC, Jensen NK and Roberson J. 2007. 'Applying DPSIR to sustainable development', *International Journal of Sustainable Development and World Ecology* 14: 543-555.

Environmental Affairs Department (EAD). 2000. *Lake Chilwa Wetland State of the Environment.* Government of Malawi, Lilongwe, Malawi.

Jamu DM, Delaney L and Campbell CE. 2006. *Transboundary Management Plan for the Lake Chilwa Catchment Area.* WorldFish Centre, National Aquaculture Centre, Zomba, Malawi.

Kalk M, McLachlan, AJ and Howard-Williams C (eds). 1979. *Lake Chilwa: Studies of Change in a Tropical Ecosystem*, Monogralhiae Biologicae, The Hague Publishers, London.

Njaya FJ. 2002b. 'Review of management measures for Lake Chilwa.' *Project paper submitted to the United Nations University*. UNU, Reykjavik, Iceland.

National Statistical Office (NSO). 2008. *Malawi Population and Housing Census.* Government of Malawi, Zomba, Malawi.

Ohlsson L. 1995. *Hydropolitics: Conflicts over Water as a Development Constraint.* Zed Publications, London.

Overseas Development Group (ODG). 2002a. *LADDER Village Report No. 26 Katanda Village, Malawi.* University of East Anglia, Norwich.

Wilson JGM and van Zegeren K. 1998. 'The birds of Lake Chilwa'pp 91-108, in: Van Zegeren K and Munyenyembe MP (eds) *The Lake Chilwa Environment: A report of the 1996 Ramsar Site study.* Department of Biology, Chancellor College, Zomba, Malawi.

Wilson JGM. 1999. 'The waterfowl of Lake Chilwa and their utilisation by local communities and conservation measures as required by the RAMSAR Convention.' *State of the Environment Study No. 20, Lake Chilwa Wetland Project.* DANIDA/Chancellor College, Zomba, Malawi.

Chapter 10

DEFRAGMENTING MANAGEMENT OF THE LAKE CHILWA BASIN RESOURCES

Peter MVULA, FRIDAY NJAYA AND MEYA KALINDEKAFE

Based on the reviews made in the preceding chapters, it is clear that there has been an evolution of management systems of the Lake Chilwa basin commons since the pre-colonial era. The management regime that prevailed in the pre-colonial era is characterised as being defragmented and centralised at the local level. The centralised system, without involvement of the resource users in the resource management was based along ethnic or chieftainship lines whereby traditional leaders had authority and control over resource management. Generally, this management system viewed the commons within Lake Chilwa basin in a holistic manner without being specific to particular resource sectors.

At independence in 1964, the question of the sustainability of the resource and capacity of the traditional leaders to control access to the resources brought in a different management regime, which emphasised particular sectors managing particular resources. The management of the commons was thus fragmented but still centralised. The resources were *de facto* state-owned in a single party government system. It was in the interest of the government therefore to manage the resources on behalf of the Malawians. For effective control, policies were developed and several programmes introduced to manage the resources in a centralised system. In this case, the resource users were not involved in the management of the commons in the same manner as was they were during the pre-colonial era where the traditional leaders had authority and control over the management of the resources.

The new government that came to power in 1994 following a return to multiparty democracy, emphasised community participation in development programmes. This called for decentralisation but still fragmented management system. In some sectors such as forestry and fisheries, co-management arrangements were introduced; with a view to change the tenure systems from state-controlled to community-owned. Subsequently, several functions were devolved to the district councils and community level committee structures that included Beach Village Committees, Area Development Committees (ADCs) and Village Development Committees (VDCs).

Despite introduction of governance reforms in the 1990s with decentralisation framework in particular, degradation of biodiversity and poverty persist. Additionally, conflicts on natural resource use prevail in the sectors. The reviews show that some policies are in conflict with each other. Specifically, the conflicts relate to management of resources in the following sectors: fisheries, land, water, agriculture, irrigation and forestry (timber and fuel wood).

This book, therefore, recommends an integrated commons management, which is decentralised and defragmented. This demands an interdiscplianry paradigm and approach top the management of the resources (Berkes *et al.* 1998, Berkes *et al.* 2003). Future research on socil-ecological complex ecosystm will need to recognise adaptive governance of social-ecological systems and investigate social sources of renewal and reorganisation (Folke *et al.* 2005). As noted by Ostrom (2009) there will be a need to gather adequate data about the characteristics of each of subsystems at various levels and understand the functioning of each SES and the basis of management actions taken.

Common property regimes are necessary especially in the Lake Chilwa basin, with recognition of the capacity of the local communities in managing the resources. It is also important to formulate conflict resolution measures. This will demand harmonisation of policies with clear definition of roles of user committees in various sectors. The devolution of functions from the central government to the local government requires political will.

Strategies on limiting access to the Lake Chilwa basin resources to prevent overexploitation, sustainable economic development for poverty reduction and conflict management address the identified problems in the preceding chapters. However, knowledge used in setting limits in the exploitation levels of the commons would require input from both natural and social scientists just as Folke *et al.* (2005) also found out. It is important to emphasie the formation of networks as already recommened in this volume. Further research work is needed to generate more information on social, economic, institutional and ecological dimension based on the DPSIR framework, which has all needed attributed to examine the Lake Chilwa basin by applying the SES approach (Ostrom 2009). Transboundary issues between Malawi and Mozanmbique that share the Lake Chilwa basin should be considred when dealing with the socio-ecological issues. Central to this is the application of formal and informal institutions.

REFERENCES

Berkes F and Folke F (eds). 1998. *Linking Social and Ecological Systems. Management Practices and Social Mechanisms for Building Resilience.* Cambridge University Press, Cambridge.

Berkes F, Colding J and Folke C (eds). 2003. *Navigating Social-ecological Systems: Building Resilience for Complexity and Change.* Cambridge University Press, Cambridge.

Folke C, Hahn T, Olsson P and Norberg J. 2005. 'Adaptive governance of social-ecological systems', *Annual Review of Environment and Resources* 30(1): 441-473.

Norberg J and Cumming GS (eds). 2008. *Complexity Theory for a Sustainable Future.* Columbia University Press, New York.

Ostrom E. 2009. 'A general framework for analysing sustainability of social-ecological systems', *Science* 325(5939): 419-422.

Defragmenting African Resource Management (DARMA)

edited by Dr. Mafaniso Hara (Institute for Poverty, Land and Agrarian Studies, University of the Western Cape) and Prof. Dr. Tobias Haller (Universität Bern)

Lindah Mhlanga; Kefasi Nyikahadzoi; Tobias Haller (Eds.)
Fragmentation of Natural Resources Management
Experiences from Lake Kariba
This book is an output of the Defragmenting African Resources Management (DARMA) Project. Lake Kariba and its environs are a complex transboundary ecosystem with multiple, competing and often contested resource uses. The individual chapters in the book describe the current sectoral approach to natural resource management, the interconnections, and the need to adopt an ecosystem approach. The book is an essential reading for students, researchers, academics and managers in the environmental and natural resources sectors interested in advancing the ecosystems approach in management of commons in general and Southern Africa in particular.
vol. 2, 2013, ca. 160 pp., ca. 29,90 €, br., ISBN-CH 978-3-643-90400-3

LIT Verlag Berlin – Münster – Wien – Zürich – London

Auslieferung Deutschland / Österreich / Schweiz: siehe Impressumsseite